GREAT

1,700 KILOMETRES ACROSS

HIMALAYA

THE ROOF OF THE WORLD

TRAIL

GREAT

1,700 KILOMETRES ACROSS

HIMALAYA

THE ROOF OF THE WORLD

TRAIL

GERDA PAULER

bâton wicks

Bâton Wicks, Sheffield, UK
batonwicks.com

GREAT HIMALAYA TRAIL
GERDA PAULER

First published in 2013 by Bâton Wicks.

Bâton Wicks
Crescent House 228 Psalter Lane Sheffield S11 8UT UK.
batonwicks.com

All photographs by Gerda Pauler unless otherwise credited.

Front cover: *Mount Makalu*
Back cover: *(Left)* Bridge on the way to Kangchenjunga Base Camp; *(Right)* Way to Tashi Labsta.

This book is a work of non-fiction based on the life, experiences and recollections of Gerda Pauler.
In some limited cases the names of people, places, dates and sequences or the detail of events have
been changed solely to protect the privacy of others. The author has stated to the publishers that,
except in such minor respects not affecting the substantial accuracy of the work, the contents
of the book are true.

A CIP catalogue record for this book is available from the British Library.

ISBN: 978-1-898573-89-0

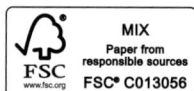

MIX
Paper from
responsible sources
FSC
www.fsc.org
FSC® C013056

Designed and typeset in TheSans and Adobe Garamond Pro by Jane Beagley, Vertebrate Graphics Ltd.
v-graphics.co.uk

Printed and bound in the UK by T.J. International Ltd, Padstow, Cornwall.

CONTENTS

Foreword by Sir Chris Bonington 7

How it started 9

A Charity Walk 13

ONE   Kangchenjunga 15

TWO   Makalu 43

THREE   Solu-Khumbu 71

FOUR   Rolwaling 81

FIVE   Helambu/Langtang 101

SIX   Ganesh Himal 120

SEVEN   Manaslu 132

EIGHT   Annapurna/Mustang 142

NINE   Upper Dolpo/Mugu 163

TEN   Jumla/Humla 197

Acknowledgements 215

Further reading 216

FOREWORD

Virtually on an impulse Gerda Pauler planned, organised and completed a 'walk of a lifetime'. At the age of fifty-five, she set off from Kathmandu, by bus, to Taplejung, walked to the Kangchenjunga Base Camp, and from there started a 1,750 kilometre-long walk along the high-level route known today as the *Great Himalaya Trail*. Sometimes with Nepali support staff, much of it alone, she crossed two snow-covered passes over 6,000 metres and sixteen more between 5,000 and 6,000 metres high. She experienced cold, wind, snow storms, rain and extreme heat – and geographical variation from jungle, through alpine landscapes, to snow, ice and wind-swept high mountain deserts. The walk took four months to complete.

A Nepal veteran with a number of visits, she says: 'I came for the mountains but I return for the people'. So too here; she met a cross section of the people who live in this Himalayan land we all love so dearly but more than that, she started on the United Nations' *World Autism Day* and used the trip to raise awareness around the subject of autism in Nepal. Autism is a difficult and demanding condition for those who suffer from it, and for their families; even more so in a developing country like Nepal.

Gerda gave up her job, met all her own expenses, and along the way (via the internet) raised over seven Nepali Lakh (9,000 USD) – enough to finance the training of two Nepali health workers in the field of autism. Once trained these two health workers, together with Autism Care Nepal, will provide training and support for other social, health and education personnel in this field. Gerda's organising ability and strength of determination, along with her dedication to a cause and her low profile manner, have always impressed me. Her book is not about her, but about this trip, across the roof of the world, done both for her own pleasure and for those who suffer from autism.

Buy the book, read it, ponder on her words, and – like her – enjoy the walk along the *Great Himalaya Trail*.

Sir Chris Bonington

HOW IT STARTED

I have been asked many times where the somewhat crazy idea of walking 1,700 kilometres across Nepal came from. All I can say is that I do not remember, but I suppose that different components came together and this plan was the result.

In 2011 the Chinese authorities closed Tibet once again and my planned cycle trip from Tibet to Nepal had to be cancelled. I was incredibly frustrated. What to do? Where to go? I had no idea. Still frustrated, I started surfing the wide and mystic world of the internet. Links led to new links, and I ended up on a page mentioning the *Great Himalaya Trail* (GHT). To me it sounded like The Holy Grail and Shangri-La all rolled in to one, and promised the adventure of a life time.

Several hours later, I switched off the computer but all the pictures, maps, blogs, articles and links had found their way on to the long term memory disc of my brain… And, as the 'delete' button was out of order, this could only mean one thing: attention! Potential danger for addiction ahead! I had experienced similar situations before and knew it was already too late. But in order to answer the question, 'How did it start?', I have to go back a long way in time – a couple of decades.

I was born in a small village outside Munich in the late '50s; an only child. Before I started school, my mountain-crazy parents would take me to the Alps – whether I wanted to or not. To be honest, I never wanted to, and all I can remember are the blisters caused by heavy boots, allergic itching caused by scratchy woollen socks, traditional knee pants that never fitted properly and my red anorak – a colour I hated as a child. Any reason to stay at home was welcomed with enthusiasm and so I developed a great talent for excuses. Going to the mountains was definitely not my cup of tea; I just did not see the point of it. Occasionally, my cousin Reinhold joined us on our holiday trips. Then at least it was less boring, but still these trips were physically demanding – which I did not like at all.

At school, my performances at gym lessons were dismal. I was the last one to cross the finish line when we did 100-metre sprints. It was me who stumbled over most of the hurdles and I assume that I resembled a heavy sack of potatoes when doing exercises on the horizontal bar. I simply seemed to lack all the necessary prerequisites for any kind of sport. And now, almost forty years later, I'm toying with the idea of crossing Nepal. It's almost presumptuous.

In general, I was more the dreamer – and still am – who got lost in time and space when reading books. But contrary to my female classmates, I systematically stayed away from all the pinkish-coloured books for girls. My heart and my soul called for adventure, suspense, exploration and the unknown wilderness. My heroines and heroes were Alexandra David-Neel, Gertrude Bell, Sven Hedin and Heinrich Harrer. They were the ones who provided new ideas for new dreams.

In 1973, I received a very special book as a birthday present. *Traumland Nepal* (Dreamland Nepal). Within a few minutes I was engrossed in the fascinating world of enchanted lakes, hidden valleys and a mystic culture that was completely unknown to me. All of a sudden I had a new dream: Nepal. I wanted to go there; at any cost. One year later, at the age of seventeen, I was determined to travel to Nepal. There were 'regular' busses running between London and Kathmandu, and the legendary overland 'Hippy Trail' was a natural source of fascination and envy to many young adventurous people in the West. My motivation was the allure of exotic countries like Turkey, Iran, Afghanistan, Pakistan, India and Nepal. One only had to buy a ticket, jump onto the next bus and a few weeks later one would arrive in Nepal. Of course, I needed some money and thus looking after babies and small children, helping younger pupils with their homework, walking elderly ladies' dogs and delivering newspapers in the middle of the night took priority over studying. As a result, I had severe problems at school and finally had to stay back one year. My parent's 'NO' was clear and candid.

In 1978, one year after my Abitur (high school examinations) Soviet troops marched into Afghanistan and the scheduled buses were cancelled for good. The dream of great adventures had come to an abrupt end. Years passed, and my life had become pretty normal; building up a professional career and getting married. Was this it? Was this my future? What about all the dreams of adventure, suspense and challenge? The marriage ended and, after having signed the papers, I more-or-less walked directly from the court to the nearest travel office and bought a ticket to Kathmandu, Nepal.

A friend, who had been to Nepal many times before, suggested a short trek and helped me to plan my trip. 'Well, why not? I will still have enough time for some culture', I thought and borrowed the necessary equipment from him. Weeks later I stood on top of Poon Hill, north of Pokhara, and upon seeing the breathtaking beauty of the high mountains right in front of me, changed my plans. My culture trip became my first trekking tour.

Since then I have visited the Himalaya about fifteen times. I have been back to Nepal, but also to North India and Pakistan. In between these trips I roamed the mountains of Europe, Central Asia and South America, sometimes by bike.

The Great Himalaya Trail was supposed to be my farewell trip; my last big adventure. But, as ever, plans are there to be changed…

A CHARITY WALK

In December 2011, I was sitting in my favourite café in Kathmandu, skimming through a local newspaper when an article about autism caught my attention. Since moving to Norway in 2007, I have worked with autistic pupils at a special school, so it did not take long to understand that there are significant differences between the conditions autistic people face in Nepal and those they meet in Norway.

According to international statistics, one in a thousand people suffer from severe autism needing intensive care and special training. Nepal has a population of about thirty million inhabitants and thus we have to assume that 30,000 autistic people live there – the majority are without any form of professional help whatsoever. Nepalese universities and colleges do not offer special education programmes providing the necessary knowledge to deal with a complex disorder like autism, and additional subjects like music and art therapy, or physio and ergo therapy are unknown in Nepal. As a result, it is almost impossible to find doctors, health personnel or psychologists who are able to diagnose the condition.

In 2008, a small group of Nepalese parents, whose children had been diagnosed with autism by specialists from abroad, founded the organisation *Autism Care Nepal*. They established a school for autistic children in Kathmandu and began to organise courses for teachers, psychologists, doctors and people interested in the field. In addition, they built up a consultation system to help other parents. The government, meanwhile, takes no responsibility and provides no funding for *Autism Care Nepal*.

Only two minutes after I had finished reading the article, I had a plan set in my mind and ready to implement. Soon afterwards launched the first version of my website on the internet: *www.greathimalayatrail-charitywalk.com* The plan was simple; I would embark on a charity walk – a long one. Having visited Nepal many times in the past and fallen in love with the country, the choice of walk was an easy one: the Great Himalaya Trail.

The GHT is a network of existing trails which together form one of the longest and highest walking trails in the world. The 1,700-kilometre Nepal section begins near Kangchenjunga on the eastern border and heads west, navigating the domains of eight of the world's fourteen 8,000 metre peaks. The route offers incredible diversity in terms of landscapes, flora and fauna, people and culture: from snow leopards to red pandas; from sub-tropical jungle to fragile high-altitude eco-systems; from the famous Sherpas, to Shamanism, to the ancient Bön Buddhist culture found still in Dolpa.

Along the way, I would talk to people I met – to teachers, to health professionals – and try to raise their awareness of autism. Through my website, I would publicise my walk and attempt to raise funds to support *Autism Care Nepal's* educational programs for teachers, administrators working with autistic children, and the families of those affected.

Of course, a Charity Walk needs a patron and when Sir Chris Bonington replied to my request within a couple of hours I was full of joy: he would give his name. His support meant a lot to me, and I deemed it a great honour to have him 'on my side'.

Sir Chris Bonington.
Photo: *Stuart Walker, Chris Bonington Picture Library*

ONE

KANGCHENJUNGA

At Kangchenjunga Base Camp

DAY 1
TAPLEJUNG – MITLUNG
Mind games

After an excruciating two-day journey from Kathmandu by bus and jeep, my guide Temba and I arrived in Taplejung late last night. This morning, when he knocks relentlessly at the door of my room at seven o'clock, I need almost fifteen minutes to sort out my thoughts. All of a sudden, I remember the radio interview I am invited to give. My journey across Nepal is not a just personal adventure, but first of all a charity walk for the Nepalese organization *Autism Care Nepal*, working with autistic children. I jump out of bed, pack my belongings in a hurry and dash downstairs for breakfast. Two journalists from the local radio station turn up before I finish the last pancake and, still munching, I answer the first questions. They want to support my project, which includes building knowledge and awareness about autism, by broadcasting general background information about this disorder. I deem this a good start to my long journey across Nepal.

After the interview, Temba and I stroll through the busy centre of Taplejung. The streets are crowded with a colourful, ethnic mixture of people who are here to do their shopping: food, furniture, electronics, or photovoltaic panels (solar cells), cooking pots, computers, fashion articles and traditional clothing. Shops and stalls offer virtually everything people need in life, and this makes the town a perfect place for my personal last minute shopping: plastic sandals, a spare torch and a nail file. The next town where I will (probably) have an opportunity to purchase items like these is Namche Bazaar; a forty-day walk from here.

'Taplejung' takes its name from the words *Taple* and *Jung*. Taple was a medieval king of the Limbu people (who originally lived in the vicinity of Lhasa, Tibet, and in the Chinese district Yunan), and *Jung* means 'fortress' in Limbu language. Thus, Taplejung can be translated as 'King Taple's Fortress'. Nowadays, there is no fortress left, but Taplejung's role as a trading and administration centre has remained.

Leaving the last houses of Taplejung behind us, I can hardly believe my fantastic adventure has started. The last ten weeks have been filled with logistics, finding equipment sponsors, donors for *Autism Care Nepal* and contacting associations connected with autism. There had been no time left to mentally prepare for a four-month journey and, as a result, my thoughts are everywhere but here, in Nepal, today. Additionally, all sorts of worries begin to trouble me; many of which are totally irrelevant at this particular moment:

What if I sprain my ankle?

What if I break my leg?

What if one of my porters gets seriously sick?

What if I have to give up?

What if…

What if…

Thoughts about my parents, friends and my former job come and go, and I forget to take in and enjoy the beauty of the hilly landscape, the exotic smells and the friendliness of the people I meet on my way to Mitlung. It feels as if I am sitting in a huge glass ball, cut off from reality, a soap bubble floating through the air.

DAY 2
MITLUNG – TAPLETOK
Puja

The trail runs gently uphill through a sub-tropical landscape with banana trees and bamboo plants. Despite it only being March, the warmth of the sun conveys the feeling of being on a summer holiday. My attention wanders from the world of thoughts to the colourful flowers, blooming rhododendron trees and bougainvillea bushes that smell somewhat like honeysuckle. Slowly, I 'arrive'.

Not far away from Mitlung, relatives of a deceased man have performed a three-day long Hindu puja beside the trail. They had erected several makeshift bamboo huts and built an altar in the biggest one to present their offerings to the deities. The people are dressed in their best clothes and there is still a solemn and sacred atmosphere around the place. Since the puja is over, the family has time to invite us for a cup of tea; together, we sit down for a while.

Puja is a Sanskrit word meaning 'to worship', 'adore' or 'pay tribute to the divine'. It is a religious procedure performed, even if the form differs slightly, by Buddhists and Hindus alike. It is underpinned by strict rules and any traditional Hindu puja has certain components that never change: the singing of devotional hymns before the ritual starts and the seemingly endless repetition of certain mantras (holy words or short prayers) during the ceremony to awaken and appease the deities. The length and the arrangements, however, can vary depending on tradition and occasion.

Offerings like flowers, rice, milk and consecrated meals are made as signs of gratitude and deference. To prepare and open up for the presence and power of the deity or the guru, the chakra ('life' or 'energy') point of

Family celebrating a puja

the people involved in the ceremony is marked with sandalwood colours. Normally, the puja ends with a prayer, and the worshipers bow or prostrate themselves to offer homage, but sometimes there is an additional light ceremony where oil lamps, incense or small pools of camphor oil are lit on stone slabs. The food offerings, which are filled with the deity's cosmic energy at the end of a puja, are distributed amongst the guests.

Hindus perform pujas for various reasons, ranging from Puja festivals like Saraswati Puja, Kali Puja, Durga Puja, and Ganesh Chaturthi to the cleansing of private and public rooms or houses. Pujas are also performed for healing, to bless babies or newly-married couples, or at funerals.

We continue along the river Tawa and reach our planned destination for the day – the settlement of Chiruwa, at lunchtime. A kind peasant family prepares food for us and sells us cold chang (local beer, usually made of rice). It is a boon to sit down and relax for a while in the shadow of an enormous banana tree beside the house. Only ten days ago, I was among the snow-clad hills and frozen waterfalls of a Norwegian winter paradise. Here, the temperatures reach about twenty five degrees, and I feel the abrupt climate change affecting me, sapping my energy.

Since the day is still young, we decide to stroll on to Tapletok where we move into the only hotel there. Well, Temba moves into one of the hotel rooms – I choose the balcony because it is too hot inside.

DAY 3
TAPLETOK – SUKATHUM
Children as porters

With an average monthly income of 20 Euros, Nepal remains one of the poorest countries in the world. In the Kangchenjunga Conservation Area (KCA), approximately 60% of the inhabitants live below the poverty line. Many of these people are farmers, families who cannot afford to send their sons and daughters to school for more than a few years, if at all. The children have to contribute to the financial standing of the family and either work on the farm or find employment somewhere else.

We meet two brothers, Lakpa and Nabin Sherpa, on the way to Lelep. I'm curious to learn more about these two children who, apparently, walk alone, and ask Temba to translate my questions. In the beginning, both brothers are shy, and it takes time before they reply.

Temba and I learn that the family owns a small farm near Taplejung and, though the fields yield enough food, the farm work provides no cash income. Lakpa is thirteen years old and has laboured as a porter since the age of ten; it was his father's decision. His brother, Nabin, was forced to leave school after grade four. Now, at the age of eleven, he transports thirty-kilogram bags filled with rice, flour or noodles up and down the hills. When they talk about their two younger brothers who still attend school, I detect a blend of envy and sadness in their expressions, envy, because the two younger brothers can still escape from the toil at home, sadness, because they will, in all probability, have to share Lakpa's and Nabin's fate when they reach their tenth birthdays.

It may sound absurd, but Lakpa and Nabin are 'lucky'. They work for the hotel at Lelep regularly and earn about 1,750 Rupees (23 USD) for the three-day trip as porters – a not only desirable, but vital source of income for the poor family.

In Lelep, where the boys deliver their loads, we meet them again. Lelep lies at the top of a steep rise overlooking the lower valley, and from it one trail runs along the Yangma Khola to the trading town of Olang-chun Gola while another passes Ghunsa and continues to Kangchenjunga Base Camp. Entering the village, I walk straight into the paved courtyard of the hotel and the adjoining shop (a well thought-out monopoly). Local children from well-off families are playing Caramboard, and there is a lot of laughter and much joking outside the hotel.

Lakpa and Nabin watch the scene from the distance. I neither see a smile nor do I hear laughter; all I recognize in their faces is sadness and resignation. They are aware of the fact that life circumstances deprived them of a carefree childhood and of even the tiniest chance of escaping lifelong hardships.

Nepal has a law that protects children, or so it says. It is illegal to assign work to anybody younger than fifteen years old. Who enforces this law, I wonder?

DAY 4
SUKATHUM – AMJILOSA
Voulez-vous coucher avec moi... ce soir?

If I had been asked whether I wanted to share my bed, my answer would have been a plainspoken 'NO', but no one had asked me. Nevertheless, an uninvited guest decided to stay over and was to make me think of him frequently over the next couple of days. I do not talk about a man who sneaked into my sleeping bag, but a flea, or a couple of them.

Upon waking in the morning, I notice a terrible itching and know at once that a flea has found a new home. These nasty little creatures do not only attack animals such as cats, dogs, poultry and mice, but also humans. Sometimes it takes less than ten minutes for them to turn their host's hair or fur into their new place of residence. I have some experience with these pesky monsters and the precautions I take are generally considered paranoid by other people. Carpets placed on chairs or wooden benches to provide more comfort have to go, as do woollen blankets in the room, and I certainly prefer sleeping in my tent to spending a night in a basic hut. These precautions, however, do not always help because, for some unknown reason, fleas love me – an affection which I do not return!

The life cycle of a flea depends on the conditions they live in, but can reach several years. Taking into account that a female flea can lay 5,000 eggs or more in her lifetime, it dawns on me that just how impressive their rate of reproduction is. While an adult flea can live up to three years between blood meals, a newly-emerged flea can only survive for about one week without food. I wonder how many of them got a one-time chance to survive by jumping into my sleeping bag...

Some animals or human beings (I am one of them) suffer from allergic reactions caused by the fleas' saliva, but worse and more dangerous than the annoying itching is the wide range of diseases that can be transmitted. The most crucial thing for me will be to get rid of my 'guests' as soon as possible; especially since I forgot to purchase any anti-itch cream.

After leaving the lodge for Amjilosa, we cross an old, rickety suspension bridge and shortly afterwards, the trail enters a dense forest with bamboo and foliage trees. Gradually, the valley narrows and we are greeted by the deafening sound of water forcing its way through the steep and narrow gorge. Waterfalls cascade down on both sides and conversation turns to shouting at one another. In some places, the sun's rays find their way down to the bottom of the gorge and make the tiny drops of water, floating in the air above the river, look like brightly shining diamonds. In other places, small rainbows form above the spray from the cascades; a stunning natural display of colour and light. It feels like walking in an unreal world of magic.

We leave the gorge and follow a series of steep switchbacks. It is a long and arduous walk, but the spectacular landscape makes us forget the hardship. On approaching the final ascent before Amjilosa, dark rain clouds mass in the south and slowly they make their way towards us. Luckily, it is not until we have been sitting in a dry and cosy accommodation in Amjilosa for a while that the first heavy drops of rain start splattering down.

DAY 5
AMJILOSA – GYABLA
Do Buddhists eat meat?

The night had been thunderous, with storm after storm rolling over the hills. For hours on end, glares of blueish lightning lit the darkness of the night and the staccato thunder, echoing around the valley, had kept me awake for hours. Now though, that the show of nature's power is over, the cloudless blue morning sky promises a perfect day.

Today, we follow the river. Some of the slopes are steep and make me suffer and sweat profoundly. On my map, it does not look bad at all, but the scale (1:150,000) is simply too large and of little use for trekking, so every bend offers a surprise, resulting in either waves of frustration running through me at the sight of yet another climb, or feelings of elation as the trail descends ahead.

Many of the tracks which cross the Himalaya are hundreds of years old. Constructed as trade routes, connecting Tibet to other Asian countries, the local population used them to travel about, as is still the case today.

Washing intestines

Usually, the trails follow the courses of rivers and thus are often prone to either getting washed away or being buried beneath landslides. Planning and constructing a safe trail is difficult and, as a result, tracks constantly wind up and down hill as they seek to avoid potentially dangerous places. It is almost as if they search for a gentle line, never finding it.

Close to Gyabla, locals 'accidented' a cow, apparently. A boy is squatting beside a creek, washing intestines with extraordinary concentration and dedication. I assume the 'accident' happened less than an hour ago.

Many Buddhists eat meat. This fact often surprises people from the West since they associate Buddhism with vegetarianism. Siddhartha Gautama, the historic Buddha, not only ate meat himself but also allowed his disciples to consume it. Buddha's words were interpreted to suggest that the consumption of meat is acceptable if the animal is not killed for a certain person. Pragmatism or splitting hairs?

As with Christianity, there are significant differences between the original Buddhist teachings and current practices. Like Jesus, Buddha did not write down his words himself. His instructions were passed verbally to his disciples and, since any narrative is influenced by the narrator's personal ideas and perceptions, this soon led to different interpretations and contradictory doctrines. Today, it is possible to meet Buddhist monks eating meat in a monastery or Buddhists who do not eat meat at all. It may also astonish many people to hear that the Dalai Lama only recently became a vegetarian – and even then, not a strict one. Whenever he considers the vegetarian meals served at European hotels too boring he orders a meat dish!

In some areas in Nepal and Tibet, meat is essential to supplement a meagre diet that would otherwise consist of roasted barley flour and potatoes. The harsh climate of the Himalaya limits the cultivation of crops and vegetables and, in order to survive, meat becomes a necessity. Buddhist lamas, however, object to killing animals, be it slaughtering or hunting, and thus the people are left with two options: a) have the job of slaughtering meat performed by a particular group or caste of people, or b) find animals that are have unfortunately been subjected to 'accidents'…

DAY 6
GYABLA – GHUNSA
Kaleidoscope

For three days, we have been walking in the Kangchenjunga Conservation Area (KCA), and this has given us the opportunity to experience some of the variety this region offers. There are 35 villages in the KCA, and the ethnic multiplicity of the inhabitants is astonishing: Sherpas, Limbus, Rais, Gurungs, Tamangs, Sarkis, Damais, Kamis, Bahuns and Tibetans. All of them have their own language and culture.

The walk from Gyabla to Ghunsa marks a change in the scenery. Everywhere along the trail, rhododendron trees, camellias and azaleas are in full bloom. The valley widens and is almost flat in places. After all the hard climbs of the previous days, I consider this a real boon.

The enormous difference in altitude (1,200-8,500 metres) entails diverse climatic zones, and the result is unique biodiversity compared to the size of the area. A journey from Taplejung to Kangchenjunga Base Camp leads the trekker through the sub-tropics, over Alpine pasture and, last but not least, to the eternal snow and ice of the high peaks.

Those who are lucky will see red pandas, Himalayan black bears, grey wolves, musk deer, blue sheep and the legendary snow leopards. Approximately 500 different birds live in the National Park, and thirty plants are endemic to this region. Now, in March, it is time for the rhododendron trees to bloom below 3,000 metres and the slopes are an undulating red sea of flowers. Nepal is home to more than thirty species of rhododendron, but this, the blossom of the *Rhododendron arboreum*, known as Lali Gurans, is the national flower of Nepal and the best known. Extensive rhododendron forests lie in the eastern part of the country and many trekkers come here to experience these beautiful, flowering trees against the background of snow-covered mountains.

The rhododendron, however, is not just a prettily blooming tree. Its green leaves are good fodder for animals in the winter months, and its wood is used for building houses, furniture and fences and for making household items, butt stocks and tool handles. In several areas, the villagers eat the flower petals and children, in particular, enjoy the slightly sweet taste. Occasionally, people even prepare a sweet drink which tastes similar to the refreshing drinks made from the hibiscus flower in the Middle East.

Traditional medicine believes in the universal healing properties of the rhododendron and use it in the treatment of skin diseases, coughs, dysentery, jaundice, diabetes, piles, enlargement of the spleen, liver disorder, worms and a couple of other things.

On arriving in the village of Phale, we decide to stop for lunch. In 2011, this Tibetan refugee settlement was partly destroyed by a severe earthquake that registered 6.8 on the Richter Scale and killed hundreds of people, both in this area and in nearby Sikkim, India. It is sobering to see that the L-shaped school complex, built by the Kangchenjunga School Project (KSP), has not yet been repaired. The once modern school is nothing more than collapsed stone walls, broken roofs and a child's drawing flapping in the wind. The pupils had been fortunate that the earthquake occurred when they were at home in their wooden houses. I think of Kashmir where an earthquake happened during school time; whole communities lost a generation.

We continue walking up the valley, following a destroyed power line that once connected Phale to the three-year old hydro power station in Ghunsa. Somewhere between these two villages, I pass a grove of firs where a memorial plate remembers a helicopter accident where twenty six people died. They were conservation specialists from Nepal and all around the globe who had been touring the region to celebrate the Kangchenjunga Conservation Area's new local management.

In the late afternoon, the first houses of the village come into sight on the other side of the river and we arrive at the suspension bridge that, during the earthquake, was just missed by a rock the size of a house. After crossing it, we enter the quiet village that still looks similar to the place Joseph Dalton Hooker described in the *Himalayan Journals*, 1855. Well, similar apart from some hotels and an electricity line, of course.

The style of the two-storey buildings is typical for Sherpa settlements; the walls are made of rough stone bricks and wooden planks cover the roofs. Several older buildings have wooden window frames with colourfully-painted carvings and pieces of striped cloth cover the eaves on top of the windows. To protect the people inside the houses from the cold and wind entering, thick, hand-woven blankets cover the doors. We follow the main path, which is partially paved with slate tiles, all the way through the village to the far side, where our hotel is.

DAY 7
GHUNSA
Rest day

Ghunsa, situated at an altitude of around 3,500 metres, is one of the main villages in the Kangchenjunga Conservation Area. Initially, we had intended to spend a rest day in Khangpachen, higher up in the valley, but the hotels are better here, and one can experience the renowned hospitality and amiability of the Sherpas. Temba had made an excellent choice yesterday because the hotel is comfortable, and our hosts treat us like members of the family. Together, we sit in the spacious kitchen; the 'heart' of any Sherpa home. Wood is burning in the clay oven, and the crackling sounds of the fire make me feel at home. A shiny, meticulously-polished pot covers one of two openings on the work surface while flames lick out at the other one. An elderly woman who tends the fire draws a branch out of the hearth again and places it beside the oven. Firewood is scarce and needs to be used with care.

I am relieved that there is already a chimney for the smoke to be drawn out of the room. The black, sooty ceiling, however, tells me that it has not been there all the time, and this evokes memories of my first trips to Nepal (more than twenty five years ago) when I regularly dashed out of kitchens, desperately gasping for fresh air.

The hotel owner's daughter-in-law shares the fate of thousands of Nepalese women; she is in charge of the kitchen. She works as a teacher at the local school and her husband left the village to seek employment. 'He flew to South Korea more than two years ago, and I have been without news from him for several months', she tells me, without any sign of worry or sadness. Traditionally, Nepalese men have always travelled due to the extensive trading between Tibet and India and are sometimes away from their families for as many as six months a year. In the Buddhist culture, this fact has empowered the women and strengthened their independence.

The young lady's excellent command of English facilitates the conversation, and I get answers to all my questions about life in Ghunsa. Winters, I am told, are extremely cold because of the altitude and thus the village is not suitable for permanent residence. All the inhabitants spend approximately two months a year (December/January) in settlements further down the valley. All the inhabitants? No, not all of them. A group of police officers have to stay behind. What for? Is there any risk for burglary? No, but Tibet is close, and people try to cross the nearby border to escape the Chinese oppression.

The policemen are not particularly keen on doing duty in Ghunsa, controlling the border year round. Life is hard, winters are cold, and heating is inefficient, if it exists at all. The biggest problem, however, are the high costs of living in this remote place. Goods need to be carried from Taplejung and over the mountains – a journey taking at least five days and for which porters charge approximately one USD per kilogram. In Ghunsa, many articles sell for a price three to five times higher than in Kathmandu. Of course, the government acknowledges the policemen's 'hardships' by paying a double wage, but even that is not sufficient to keep up the standard of living they are used to in the capital city. As a result, they visit local shopkeepers and hotel owners regularly and demand 'support' in the form of money, alcohol or food. Fearing problems, the villagers dare not turn down the guests in uniforms.

DAY 8
GHUNSA – KHANGPACHEN
Hotel Yak

Last night, a fierce thunderstorm raged through the valley and brought rain, hail and snow. However, the warm rays of the sun hit the wide valley early in the morning, and the snow is melting quickly.

Changing a plough blade

The one-day break had done me good. I feel well rested, and I am sure to reach Kangchenjunga Base Camp, the official starting point of my Nepal traverse, on April 2nd: the United Nations' World *Autism Awareness Day*. Today, we will walk on to the settlement of Khangpachen, which lies about 700 metres higher than Ghunsa.

The trail runs through an Alpine landscape that is still in a bear-like hibernation. The meadows are brown and look dreary. A travel brochure, describing this area, talks about a great variety of wild flowers and I try to imagine what these pastures will look like in the summer: primroses, daisies, poppies and edelweiss. I suppose it will take some time before summer arrives here, turning the meadows into a paradise for botanists.

There are small farms along the way, and everywhere people are busily ploughing the barren fields. With no machines to help them, this is time-consuming and backbreaking work. Dzos (infertile male crossbreeds between domestic cattle and yak) are harnessed to the ploughs to draw shallow furrows in the hard and stony soil and soon the people will sow barley or plant potatoes.

Beside one of the small farms, people have gathered round. When coming closer, I see that the farmer's plough blade is broken. Even nowadays, most of the equipment is still hand-made and can thus be repaired easily. After spending some time discussing the problem, the damaged plough blade is taken off and a new one fastened on. Helping each other is an integral part of everyday life in the mountains and goes without saying.

On arriving in Khangpachen, I am disappointed. For some unknown reason I had expected to find a proper village here; do not ask me why. Of course, there are some houses, but they are only used in the summer by herders who come to this area with their animals. Right now, all the buildings but one are abandoned and thus I am even more surprised to meet another tourist at the Hotel Yak. Heinz, an engineer from Switzerland, has just returned from Kangchenjunga Base Camp. This presents an excellent opportunity to get the latest information about the conditions of the trail higher up. We learn about a landslide blocking the way between Khangpachen and Lhonak. According to Heinz, the scree is loose and awkward to move on and the stones and rocks are wet and slippery. 'Take good care', he advises us, 'try to cross this landslide as early as possible. The danger increases the later you get there'. Well, I fear this can only mean one thing; an early start.

DAY 9
KHANGPACHEN – LHONAK
Where is water?

The ascent to Lhonak is surprisingly easy. The trail is good and, apart from one steep section, we gain height gently and gradually. Yet, we still have to cross the landslide Heinz had talked about yesterday. This turns out to be a challenge. Some of the blocks, hanging above us, seem ready to come down at any moment, and I can feel my shoulder muscles tense. With the rocks merely biding their time, this is not a place to linger and there is one rule only: 'walk on and do not stop!' The sand slides away under my mountain boots as I look down into the abyss beside me. It comes as a great relief when this part of the walk lies behind us.

In the early afternoon clouds accumulate and conceal the magnificent view of the snow-covered peaks. Will the weather thwart my plan? All I ever wanted for my official start was to reach Kangchenjunga Base Camp in bright sunshine and taking pleasure in an unobstructed view of the third highest mountain in the world. Did I want too much?

Like Khangpachen, Lhonak is not more than a tiny settlement consisting of a few scattered stone houses and huts used by shepherds in the summer. One of these serves as a hotel, but right now it is still closed. Luckily, we knew this would be the case and Temba has collected the key from the owner's relatives in Ghunsa, who told us that there is food in the kitchen. Nothing can go wrong, I assume.

Even before the sun disappears behind the summits the temperature plummets. It is ice cold up here at an altitude of almost 4,800 metres. We start a fire to warm up the room, prepare a meal and, most importantly, to make tea. The big problem, however, is that we cannot find any fresh drinking water. Not one single drop comes out of the pipe behind the house. Everything is frozen. There is only a tiny amount of snow covering the ground between the small buildings and even that is pretty worthless – definitely not good enough to be used for melting and drinking. We search the hotel and, finally, discover a large canister in one of the storage rooms behind the kitchen. It does not contain much water, but it will be enough for two days. We do not give the water quality a second thought…

DAY 10
LHONAK – KANGCHENJUNGA BC – KHANGPACHEN
Dehydrated

Since we intend to walk up to Kangchenjunga Base Camp and then return to Lhonak and the 'hotel' for another night, we start early, taking only some food, drink and the camera with us. The rest of our luggage is left behind.

The weather could not be better. Bright sunshine accompanies us all the way and the sun-bathed flanks of Gimmigela, Wedge Peak, Nepal Peak and Tent Peak present the most superb scenery for photographers. Four hours later, we arrive at the area which is defined as Kangchenjunga Base Camp, at an altitude of 5,140 metres. Kangchenjunga towers up another three and a half thousand metres into the blue sky above us.

Located along the India-Nepal border, Kangchenjunga is, at 8,586 metres, the third highest mountain in the world and until 1852 was thought to actually be the highest. As early as 1849, the Great Trigonometric Survey of India had come to the conclusion that Mount Everest (known as Peak XV at the time) was the highest peak, but it was not until 1856, after further verification of all their calculations had been made, that it was officially announced that Kangchenjunga was 'only' the third highest mountain in the world. It was first climbed on May 25th 1955 by two members of a British expedition: Joe Brown and George Band. Out of respect for the belief of the local people in Sikkim, who hold the summit sacred, they stopped a few feet below the highest point. Since then, many mountaineers have followed this tradition.

Since one part of the Kangchenjunga massif lies in Sikkim, India, and the other one in Nepal where people speak Tibetan, the name of the mountain originates in two languages. In the Tibetan language, *Kang* means 'snow', *Chen* 'great', *Ju* 'treasure' and *Nga* 'five' – with the name Kangchenjunga meaning 'the five treasures of the great snows'. Whether this refers to the five summits, the five glaciers or the five repositories of God, is not certain. But the latter interpretation is the most common one, assuming that the five treasures represent gold, silver, gems, grain, and religious books.

Deriving the name Kangchenjunga from Sanskrit gives another translation. *Kanchana* means 'gold', and *Ganga* is 'the river which flows in the region'. The river shines like gold and hence this mountain received the name Kanchana Ganga.

Be that as it may, the mountain has a majestic peculiarity around it, and I respect every mountaineer who climbs it and comes back down safely. Many of them, including Reinhold Messner, describe Kangchenjunga as the most dangerous and one of the most difficult 8,000 metre peaks in the world.

However, my happiness on seeing this giant of rock, ice and snow in perfect weather is somewhat clouded. Since leaving the hotel this morning, I have not felt well. Mild stomach cramps spread to my kidneys on the walk to Base Camp. Of course, I wonder what causes the pain, but cannot find an explanation that makes sense. What is most odd is the fact that whenever I take a sip from my tea I feel a strong reluctance. I cannot swallow one single drop. It is as if my entire body cries out 'NOOOOO'. On the one hand, I am convinced that listening to the body is always the best idea but, on the other, I know that drinking a lot at higher altitude is of utmost importance in order to avoid high altitude sickness. I am trapped in a quandary, but in the end I opt against drinking.

When walking back to Lhonak, Temba begins to suffer from similar symptoms, and we suspect the quality of the water is the cause of our problems. Back at the hotel in the afternoon, we begin to search for drinks feverishly. We open all the wooden boxes, look under the beds, empty the plastic barrels, examine the shelves… we practically turn everything inside out, but do not find one single bottle of lemonade, not one can of Coke. After a short 'crisis conference' we decide to walk on to Khangpachen.

After a twelve-hour day without drinking, we finally arrive at Hotel Yak, exhausted and feeling like parched prunes. We drink one pot of hot tea at a time to balance the loss of liquid and hope for the best.

DAY 11
KHANGPACHEN – GHUNSA
Cookies in Ghunsa

Around noon, we get to the village of Ghunsa, bringing with us a thunderstorm that opens up the heavens with rolling thunder, hail and snow. Suddenly, I perceive a craving for a sweet 'mood enhancer' (I put the blame on the weather) so, whilst Temba goes on to the hotel we had stayed in before, I walk down the muddy 'Main Street' in search of the local shopping centre.

All I discover is a tiny shop, which is part of a private house. To my disappointment, the entrance is closed, but giving up is not my thing. I walk to the back of the building and, when opening a door there, I find myself in the owner's kitchen and several people squatting in front of a clay oven. No one seems to be surprised by a total stranger wandering into their kitchen. My Nepali vocabulary is somewhat limited, but I am able to explain what I want: 'Some sweet biscuits'.

The lady shopkeeper gets up and, waving her hand, asks me to follow her. She smiles and goes to the shelves at the back of the room where an array of products in all shapes and sizes wait for customers. She takes down a dusty package with Chinese writing on it, but the dull cover makes me doubt that there are biscuits inside, and so I turn the package round again and again, hoping to find some English words.

I try to hide my scepticism when asking, once again, if the package really contains biscuits; it is always good to be on the safe side. She smiles, nods and repeats my words. Still, I am not convinced, but more questioning would appear impolite and so I pay the price she asks, thank her and leave the shop.

Outside, I still wonder what I have bought and unpack the main package only to find a number of small packages inside; each one individually wrapped in plastic. Opening one, I see a grainy brick that looks similar to the ones we use to start fires with back home.

I hesitate before I bite into it, looking around to see if anyone is witnessing my attempt to eat a fire brick. It is hard, grainy and tastes like heaven – truly delicious! It reminds me of a Swiss speciality – Schweizer Nusstängeli.

Later Temba laughs at me when I tell him about my doubts. He calls them 'Army Cookies' – bought by the Nepali Army from the Chinese for their troops – and popular with traders crossing the border to Tibet. 'Nice to have when life is hard', Temba says with a broad smile.

As planned, Lakpa and Sonam have arrived in Ghunsa with some more equipment: tents, kitchen utensils and food. Both men are Temba's relatives, and thus I know I can count on them. They will help us to walk on to the Makalu region tomorrow.

DAY 12
GHUNSA – GYABLA
Good advice

When deciding on a guide, I want to be sure that he (or she) will be self-reliant, able to make the right decisions at the right times and without hesitating. Temba and I know each other from a previous trip, where he showed qualities that I appreciate in a guide. He is honest, shows maturity and not only cares for my well-being but also for that of the staff and people we meet along the way. Today, Temba proves to be an excellent guide once again.

During the last ten nights a lot of new snow fell at high altitude and the likelihood for avalanches increased significantly. Is there a chance to avoid crossing the Nango La (4,820 metres)? The map does not show any alternative routes to Olangchun Gola, but this does not necessarily mean that they do not exist. Early in the morning, Temba virtually walks from house to house to ask locals about other trails. It turns out that there is a track which the people from Ghunsa use when the weather conditions are poor.

One hour later, Temba returns with a vague description of the trail and a rough drawing. I take a closer look at the makeshift map consisting of a few lines, three arrows and a circle that is meant to be a lake. Is this rough sketch sufficient to guide us over the mountains? After all, it is a three-day journey to Olangchun Gola. I feel doubts coming up, but Temba is 100% sure. 'No problem, I will find the way', he says. This convinces me that he knows what he is doing. We leave Ghunsa and return almost as far as Gyabla where we 'turn to the right'.

Tonight, Lakpa, who not only joins the group as a porter but also as a cook, prepares Dhal Bhat for us for the first time. His cuisine is astonishing. What is Dhal Bhat? Usually, the dish consists of a Mount Everest-sized heap of rice served with a lentil soup that comes separately in a bowl. Lentils come in almost any colour (yellow, orange, red, green, black) and thus, every meal is a visual surprise. Depending on the season and availability, one gets different vegetable curries, omelettes or even a mushroom curry in addition. The true heroic trekker may even consider a meat or fish curry as a side dish. Admittedly, I am not that courageous.

There are two reasons why I opt for Dhal Bhat. Firstly, one gets a tremendous amount of calories and carbohydrates for a good price. Secondly, there is always a good chance to be offered a second helping. On my first trip to Nepal in the 80s, I fell for Dhal Bhat and adopted the habit of eating the national dish twice a day, like the locals. 'Boring!', some may state. No, not at all. Every cook or housewife has his or her own recipe and, therefore, the dish never tastes the same. When travelling in Nepal, I make a chart for the region I cross. Lakpa's Dhal Bhat ranks exceptionally high.

DAY 13
GYABLA – KHARKA
Lost in the jungle

My first breakfast in the tent consists of chapattis and omelettes, and hungrily I eat a whole pile of them while sitting in my sleeping bag. It is only now that I feel that my 'great adventure' has started. We pack everything, place the bags into Lakpa and Sonam's dokos (wicker baskets) and start walking through a jungle-like forest. The trail is surprisingly well-maintained and, in the beginning, we make good progress. The weather gods, however, seem to get angry about one thing or the other and send dark clouds in our direction. A few hours later, we fight against a hailstorm that, eventually, turns into a snowstorm.

We reach the lake shown on the rough drawing without any problems, but the dismal weather with thick clouds all around us makes trail finding difficult, if not almost impossible. How and where do we get down from here? In the mist, everything looks the same. The makeshift map shows an arrow pointing to the left of the lake, but we cannot detect any sign of a trail under the hail and snow covering the ground. We decide to leave the luggage behind under a big boulder, split up and search for the path. I guess that fifteen minutes have passed when, all of a sudden, Sonam cheerfully shouts 'Found it!' We set out in a single file and start the long descent. Several times we lose the trail and have to look for it time and again, but eventually reaching a forested area without snow, we begin to make better progress. Unfortunately, it doesn't last.

The fierce storms that raged through the Kangchenjunga area recently resulted in countless tree trunks blocking the path. They frequently force us to leave the trail, and, while descending through the almost impenetrable thicket, we lose orientation. In this situation, neither the map nor the drawing proves to be of any use, and all we can do is hope to get back to the trail at one point.

It feels as though I have walked through the jungle for hours on end when I, unexpectedly, find an empty Coke can. 'A clear evidence of civilization', I decide and inform the others of my extraordinary find. My experience tells me that litter is always disposed of close to tracks and my conclusion turns out to be correct. The trail is less than five metres away and we are glad that the creeping and crawling through the undergrowth is over. Yet a lot of time was lost on our jungle adventure, and we will now not make it down to the bottom of the valley before total darkness sets in. As soon as the forest lies behind us, we look for a suitable place to spend the night. This does not take long and, when an abandoned shepherds' hut comes into sight right in front of us, we out of hand declare the dilapidated building to be our home for tonight.

We have hardly settled in when rain starts to pour down again. The roof of the shelter leaks considerably, and Pimba and Sonam put my tent up inside the building in a hurry. As an additional protection against the rain, they roll out the kitchen tent across the roof.

Soon afterwards, a delicious soup is boiling in one of Lakpa's pots, and the pressure release valve of the pressure cooker makes the familiar hissing noise which tells me that it will not take long for the rice to be ready. The hard time is forgotten before having finished the first plate.

DAY 14
KHARKA – OLANGCHUN GOLA
Traders

Bright sunshine in the morning makes it easy to get up and leave early. Still, we have to descend 6-700 metres to reach the bottom of the valley. The path, running through a thick bamboo forest, is like one of these giant water slides found in Aqua Parks.

Our slide, however, is not filled with water but mud, and soon we commence a glissade down in the ankle-deep dirt. Of course, it does not take long before I 'take a ride'. Everything is covered with dirt: shoes, pants, jacket and rucksack. Luckily, it is only me who has a camera, which is safely stored away and so there will be no documentation of my limited abilities as a mud slide artist. I am relieved when we reach the bottom of the valley and continue on flat and dry ground.

Olangchun Gola is situated in the Tamur valley, and its residents came from Tibet a long time ago. The name of the village originates in a folktale about a wolf (*olang*) that showed a trader (*chun*) at this place (*gola*) a way to Tibet. The inhabitants of the village have been passionate traders travelling as far as India to purchase – amongst other items – cloth, grain,

Bamboo forest

brown sugar and cigarettes which are freighted on the back of yaks to Tibet. There, the commodities were bartered for salt, wool and carpets.

At first glance, the village appears to be from another time, another century. Yet, due to extensive trading and travelling, the inhabitants of Olangchun Gola are reasonably well off and can afford to send their children to schools in Taplejung – or even Kathmandu – and they are well-informed about what is going on in the world. While sitting in front of the hotel and waiting for a cup of tea, I see a young man walking up to me. After exchanging small talk, he asks, 'Is it true that you are from Norway?' Usually, this question alone comes as a surprise, because not many Nepalese people know that a country called Norway exists. This young man, however, has an even bigger surprise in store: 'Isn't it terrible what happened last summer in Sundvollen? Anders Behring Breivik, who killed all these young people and caused so much pain to hundreds of family members, relatives and friends, must be a totally mad man'. I am completely speechless because being confronted with the, probably, darkest day in Norwegian history in a faraway mountain village in Nepal is beyond anything one would expect. For a while, we sit there together and talk about Sundvollen.

We set up our tents beside the only hotel in the village and spend the evening in our hosts' kitchen, looking at the pictures I had taken on the way from Taplejung. Their two children know many of the people shown in the pictures and their merry laughter fills the kitchen when they recognize some of their friends from school.

DAY 15
OLANGCHUN GOLA – KHARKA
Visitors for Lunch

The ancient trading route to Tibet, running through Olangchun Gola heads north. To reach our destination, the village of Thudam, we have to follow a trail over the Lumbha Sambha La which lies to the west. Hardly anybody takes this route because it is much easier to walk down the Tamur valley for trading than to cross a 5,000 metre pass. Given the unfavourable weather conditions, it seemed sensible to find a local guide who would be familiar with this route. After some negotiation, our host agrees to join us for two days and so today we set out together.

The trail starts with a moderate climb, which I thoroughly enjoy. After having followed the wide valley for a couple of hours, we take a break beside a creek, and while Lakpa prepares the meal, I recline against a rock. The bubbling noise of the water is like soothing, meditative music that makes me doze off and escape to a world of comfortable armchairs and soft pillows. All too soon, however, fragrant clouds smelling of herbs and spices reach me and bring me back from the land of dreams. It is only then that I realise how hungry I am and look forward to a palatable supply of calories. But the smell also attracts some uninvited guests. Yaks.

My relationship with yaks (male) and dris (female) is ambivalent. On the one hand, I adore the grace and agility they show when moving up and down steep slopes; on the other, their size and the vicious-looking horns scare me. An adult male yak may weigh more than one ton, the height at the withers can reach almost two metres and the horns are up to one metre long. Dris are much smaller, but even they can weigh 500 kilograms. Domesticated yaks are often described as aggressive and wild but at the same time, are considered extremely shy and nervous. Even their shepherds tend to keep their distance in order not to scare them or even trigger off a stampede. They know that it never takes long for these gentle, doormat-like creatures to become wild beasts, suddenly possessed of seven devils. Their clumsy appearance is deceptive...

More and more animals move closer to our lunch spot, and Sonam tries to shoo them off, without success. Some of the dris have calves, and thus they are on the alert and aggressive, shuffling their small hooves, staring at us with their round, brown eyes and lifting their tails. I fear the worst; yak attack. Desperately, I look for large boulders and huge trees that I could run to and hide, but I know that yaks can move fast; faster than me. Could I, possibly, stop one of these beasts by throwing

Dri with calf

stones at it? That's doubtful. Would playing dead be of any use? At least, it works when being attacked by a mother grizzly... I would rather not give this method a try.

Suddenly, we hear a voice shouting somewhere in the distance and soon a man emerges from the nearby forest. He turns out to be the shepherd. Slowly and warily, he approaches his animals, and after a while they calm down. He explains that the salty smell of our meal had attracted them and to make his yaks move away from us, he empties a bag of salt further up the slope. Gracefully, the yaks walk away. 'They are gone. We got rid of them', I think with great relief and concentrate on my meal.

After our lunch break, we trudge on again, slowly but steadily up the valley. Here and there, the snow has melted, disclosing patches of last year's brown grass and dwarf rhododendrons. Most of the time, however, a thick white layer covers the trail, and only the contours of the terrain tell us where the path runs. Occasional sun rays fight their way through the clouds and reveal a glimpse of the peaks lining the valley, but soon dense clouds are massing, hiding the mountain scenery behind a curtain of grey and when the first snowflakes begin their lively dance around us, we decide to stop for the night.

DAY 16
KHARKA – LUMBHA SAMBHA LA
The first 5,000 metre pass

A long and tiring section awaits us. More snow fell last night, and we suspect the trail to be buried higher up on Lumbha Sambha La. Deep, powdery snow will slow us down and make the ascent difficult and challenging. The area we pass through is complex because there are a number of valleys leading up to passes that 'feel' right but which would take us nowhere – or at least, not to Thudam. Our local guide from Olangchun Gola is excellent. With an almost somnambulistic reliability, he walks in front of us, not hesitating once at any of the numerous trails crossing. He just knows where to go.

Usually, Lakpa prepares a hot meal when we stop for lunch and we appreciate these two-hour breaks. But today, since we do not know how long it will take us to reach the pass, we only stop for twenty minutes, giving us just enough time to drink hot tea and eat a few chapattis that Lakpa had prepared yesterday evening and soon we set out again. As feared, the snow is deep, and for Temba and the local guide who walk in front of us, it is back-breaking work. Again and again they get stuck in the snow, almost up to the hips, slide down, try again… But they plough on through the soft snow, providing an opportunity for me to keep up with them, plodding up the trail they have broken.

Finally, we get our first look at the pass, and our local guide returns to his villages with the words 'You simply walk straight on from here.' We continue our struggle up the steep mountainside. As ever, we notice some murky clouds gathering round the high peaks, but at this very moment bright sunshine accompanies us and so we do not worry. The fact that Lakpa stays behind, however, gives cause for concern. Apparently, he is no longer able to keep up with us. taking more and longer rests than normally. Is he just tired or affected by the altitude?

From the very beginning, we had planned to cross the pass and descend as far as possible the same day, thus avoiding (or minimizing) the dangers of altitude sickness. We stick to this plan. Temba leaves his rucksack with us and slides back down to take Lakpa's load. However, we did not factor the weather into our plan and before Temba and Lakpa reach the pass, the murky clouds have gained strength and we are enveloped in dark, dense mist as heavy snowfall sets in. Within five minutes, the visibility is reduced to almost zero and an ice-cold storm makes breathing difficult. Proceeding under these conditions seems too risky because, according to information we had received from the local guide, there are a couple of steep gullies ending at almost vertical precipices. Good visibility is essential for a safe descent.

When Temba and Lakpa arrive, the latter shows early, but clear, signs of high altitude sickness: tiredness, exhaustion and shivering. What to do? Turn back? We are aware of the fact that due to the raging snowstorm our footprints will disappear within a minute or two. There will be no chance of finding the way back down. The only alternative that makes sense is to pitch tents, keep Lakpa warm and give him as many hot drinks as possible.

While Sonam, Pimba and I level out a campsite, Temba looks after his uncle. As soon as the tents are pitched, we place Lakpa in his sleeping bag, put some bottles filled with hot water into it, and spread all the down jackets that we do not need right now, and a spare plastic tarpaulin, over him. We only hope that this is sufficient. There is not much more we can do for him right now.

DAY 17
LUMBHA SAMBHA LA – THUDAM
High altitude sickness

Early in the morning, it is not only a brightly smiling Lakpa who greets us but also a brightly shining sun. We are happy to see that our cook is much better and that all the signs which indicated altitude sickness have disappeared.

Many people hold the erroneous belief that only trekkers and mountaineers can be affected by high altitudes; far from it. Nowadays, the majority of guides and porters no longer live in the mountains but in Kathmandu at an altitude of approximately 1,300 metres.

All guidebooks about trekking contain, at least, a short chapter about altitude sickness. My advice is to read as much as possible about altitude sickness and its different forms. Even better sources of information are experienced trekkers, mountaineers and consultation places in Kathmandu and along the popular trails. Before I left Norway, someone had asked me what I thought to be the greatest danger on the Great Himalaya Trail and feared the most. I did not have to think twice. My reply came out of hand: High Altitude Sickness.

Although I have never had any serious problems with altitude on my previous trips, I know that there is never any guarantee for a future trip.

I have seen enough trekkers being carried down by porters (who often risk their own lives to do so) or being rescued by helicopters from faraway places, to understand that the danger is omnipresent, even as low as 2,000 metres. High altitude sickness can even occur when coming down from a 6,000 metre peak simply because one probably feels excited about a successful climb and ignores basic advice. No matter if ascending or descending, I stick to a few key rules – without exceptions:

- Walk slowly but steadily.
- Avoid any fatigue.
- Keep warm.
- Drink enough hot drinks – a minimum of three to four litres a day.
- Make sure that the kidneys work properly by checking the 'output'. Storage of liquids in the tissue is one of the signs, mostly in women.
- Never compete with others.
- Have a rest day as soon as the slightest unfamiliar symptom shows: headaches, coughing, swollen face, tiredness, dizziness, loss of appetite…
- Go down if the symptom is still there the next day.
- Show responsibility for the rest of your group.

The local guide from Olangchun Gola had mentioned a four-hour descent but the reality is different. For hours on end, we plod down the steep slopes covered by deep, new snow. In the early morning, the frozen surface crust still carries the weight of a body plus a load, and occasionally we sit down and commence a glissade down a steep part of the slope. With time passing, however, the crust becomes soft and collapses under our weight. In order to avoid sinking into the snow, we begin to pull the rucksacks and dokos behind us. Unfortunately, this technique is no longer useful when we arrive at lower altitudes where the snow is soft and wet. Frequently, we break through the crust up to the thighs and descending becomes tedious work. When we finally see the first houses of Thudam far below beside the river, afternoon is already merging into evening and the sun is fighting its daily battle against dark clouds that, probably, bring more snow.

Some people consider the village one of the remotest settlements in this region. A quick glance at the map tells me that Tibet, lying to the north, is closer than any Nepalese village to the south. This explains why nobody from Thudam ever goes there for shopping or trading. Judging from what I see and hear, the government of Nepal has forgotten this hamlet and its inhabitants, and in return the locals no longer feel a part of Nepal. Hardly anybody speaks Nepali because the Ministry of Education stopped sending teachers; the local school has been closed for several years. There is neither a postal service nor a health post. The public water taps providing safe drinking water broke down years ago, but the villagers lack both the necessary money and the required knowledge to fix them.

After a less-than-rudimentary washing session, I hang out some of my clothes behind my tent to dry. Needless to say, I do it very discreetly. Some of the local women, however, discover my underwear. They gather round, giggling and laughing and, by employing gestures, they explain that they would appreciate a more minutely inspection of a few things. This ends in even more giggling. Admittedly, my sports underwear is not particularly exciting. Do they probably wear sexier apparel under the long Tibetan dresses?

DAY 18
THUDAM – KHARKA
Where is my coffee?

Last night, the temperatures fell below zero and upon crawling out of my tent, I see fresh snow covering the landscape. The small settlement of Thudam resembles a bleak black-and-white picture and my little red tent adds the only colour to the dreary scenery, which is intensified by the quietness of the early morning. What will life be like for the inhabitants in the future, I wonder? What will change and when?

There are not many things of real importance to me when travelling, but a big cup of cappuccino after breakfast is one of the best things I know. Of course, I have to reduce my expectations while trekking in Nepal and I am happy with ordinary coffee and will even drink the instant version of the black brew. That's the way it is in the Himalaya. The event-packed nature of the previous two days made me forget my addiction, but today I would appreciate a cup of strong coffee. The only question is: where is my coffee? Since our local guide left us, nobody in my group has seen the package.

Feverishly, we start searching through all the bags again, but cannot find the coffee. In the end, we come to the conclusion that it 'joined' the local guide on his journey back to Thudam. My first reaction is anger, but soon I calm down, telling myself that it is better sitting in the right valley without coffee than drinking coffee in the wrong one. I forgive him with all my heart and hope that he will enjoy my coffee.

The trail between Thudam and Chepuwa winds through a densely forested area and, in bad weather, this section is precarious. A precipice to the left tells me that it is the best to avoid any incautious movement as it may end in a fatal accident. In several places, the path is carved out from the rock, but since the height of the roof is based on Nepali people's height it is far too low for me and my 178 centimetres. Add the danger of a rucksack getting caught at a rock jag, and one is at risk of losing balance and plunging into the gorge.

Another risk factor is the amount of traffic along the way. Woodcutters are working everywhere, and porters are busy freighting cut up logs to Thudam. 'Cut up', however, means that the logs are 'only' two and a half metres long. Unfortunately, the path is rarely wider than one metre and forces the porters to walk sideways most of the time. I believe that none of them are genuinely happy when running into us on the narrow trail because any encounter requires hazardous sidesteps.

'What do they do with all the wood? Isn't it prohibited to cut trees?' I ask Temba. 'There are no controls out here in the forest, and the people deliver the logs to Tibet where they get good money for them. The local people are poor, and the income from selling yak wool and butter is not enough. They depend on these additional earnings and there is usually a chance to get some food for free. Do you remember the empty rice bags we saw outside Olangchun Gola? *Tibetan and Chinese aid for our friends in Nepal* was written on the material', he replies. Yes, I had noticed these bags several times.

After nine hours of tiring ups and downs, we are totally done and decide to pitch tents beside an abandoned shepherds' hut.

TWO

MAKALU

Lakpa and his two youngest children

DAY 19
KHARKA – CHEPUWA
Responsibility

The grey, cloudy morning sky predicts more rain. Since the walk to Chepuwa is reasonably short, we have no reason to rush and I have time to enjoy the beauty of the blooming rhododendron and azalea trees. The spring season is in full swing and the hills are swathed in red, pink, purple and white flowers. This sight truly resembles something out of The Wizard of Oz. Just before arriving in Chyamtang, where we want to stay for lunch, the slight drizzle, which has accompanied us for two hours, turns into heavy rain. Luckily, we reach the village before getting totally soaked.

Sangbu Bhoti, Temba's uncle, is the headmaster of the local school, and while waiting for our lunch, he tells us about the latest community project: a hostel for girls. Even nowadays, girls rarely attend school in rural Nepal, and when they do, many of them will not complete their schooling. The higher secondary school in Chyamtang includes a hostel for older boys living in faraway settlements but no accommodation for girls. Building a place where girls can stay overnight will hopefully change the situation in this area.

From our lunch place, we observe a group of workers digging out the construction site while heavy rain is pouring down. Sangbu tells me that almost all the local people are actively involved in the project by dedicating a few days' work to it. The costs for the material are met by an NGO that bases its work on the idea of helping people to help themselves. Watching the workers and listening to Sangbu, I am sure that the planned construction of the hostel will be carried through. The passion he radiates is seemingly affecting the inhabitants of the village, and this will make the project a lasting success.

After a good meal, Sangbu invites me to see the school. The main building houses a library that is also used as a meeting place for local people to attend lectures on hygiene, family planning, healthy nutrition and general development projects. All along the walls, there are colourful displays showing the human body and statistics. Although this is quite entertaining, my main focus is on something completely different: the solar power plant. Here, I can charge all the batteries of my electronic equipment: note-book, mobile, iPod and cameras. The prevailing chaos of dangling cables is not confidence inspiring at all, but the station turns out to function perfectly. I am looking forward to listening to music or an audio book tonight.

A leisurely stroll takes us to Chepuwa where we move into Phurbu Bhoti's house. Phurbu is one of Temba's brothers and, as a welcome drink, his family serves Tibetan Tea. It may confound the reader to learn that I actually like this rather unusual drink and cannot understand why this tea has such a dreadful reputation. Admittedly, the passionate connoisseur of tea will need some time to get used to the combination of tea, salt and butter, but as soon as one successfully manages to think of a hot bouillon, the chances are good to become a butter tea connoisseur, like me. For me, the quality of the tea as such is not of great importance. My focus is on the quality of the butter. In remote areas of the Himalaya, butter can be rancid sometimes – in which case even I consider the tea to be undrinkable.

Sangbu's Tibetan Tea is delicious, and I want my cup to be filled again and again – although Sangbu would have done this anyway as it is the custom to refill a guest's cup to the rim as soon as the first sip has been taken. If belonging to the group of people who believe Tibetan Tea to be undrinkable, this custom presents a quandary... in which case I highly recommend some training at home to get used to this traditional drink. For those in need of 'training' here is the recipe:

Boil the tea leaves in water for half a day until the liquid has achieved a dark brown colour. Skim the tea and pour it into a special tea churn cylinder together with fresh yak butter and salt. Churn everything vigorously until the liquid looks like a stew or thick oil. Pour the tea into clay teapots. Enjoy drinking!

DAY 20
CHEPUWA
Rest day with Tongba

Children's laughter from somewhere in the house brings my deep sleep to an abrupt end, and experience tells me that the children will soon appear beside my bed on the balcony. Foreign guests rarely come to Chepuwa and hence visitors from abroad invoke curiosity. I decide to crawl out of my sleeping bag and go to the kitchen where the family members are already extremely busy.

The children are not the only ones who are curious. Temba's relatives are eager to meet me. All day long we visit brothers, sisters, cousins, aunts, uncles, sisters-in-law and brothers-in-law. Needless to say, there is food and drink in abundance. According to Nepalese customs, the guest has – at the very least – to taste everything offered by the host. Given the size of a Nepalese extended family, visits can turn into hard work! Luckily, I have already spent twenty days in the mountains and I am hungry all the time.

In Chepuwa, I also learn more about Lakpa's fate. Chejik, his wife, died when giving birth to the youngest daughter. Like in many other cases, she also lost her life because there had been no doctor around. Around six per cent of mothers die during childbirth – an appallingly high figure compared with Germany, where it is about 0.1%. And according to UNICEF, eighty per cent of those that die could have been saved.

Now, Lakpa lives together with his two oldest children in Kathmandu, since schools are better there. The two youngest stay with relatives in Chepuwa. His visits are few and far between because the journey from the capital to this remote village is time consuming. Depending on road conditions, the trip can take up to ten days. Of course, his children are very happy to see him and, all day long, they tell stories to each other, and they laugh, kid and joke a lot. Who knows when daddy will come to pay them a visit next time?

In Chepuwa, I am offered Tongba for the first time on my journey. Tongba is actually the name of the wooden container, but it is the alcoholic contents of that container, prepared by cooking and fermenting whole grain millet, which I am excited to try. For the local people in

Nepal, Sikkim and Darjeeling, Tongba is what sake is for the Japanese, vodka for the Russians and wine for the French. It is not a drink only but an integral part of the culture.

Spontaneously, I develop a taste for this 'local beer' and want to know how to make it. Maybe I will invite my friends back home round for a Tongba party after returning from Nepal? The mere thought of the penetrating odour of fermenting millet filling a house where there are more rules than people makes me giggle.

Tongba preparation starts by cooking and cooling the millet. Murcha (a kind of yeast) is added and the resulting mass placed in a woven bamboo basket lined with green leaves and covered with a thick cloth. Depending on the temperature, the mass has to stay in a warm place for one to two days. Then it is packed tightly into an earthenware pot or plastic jar which has to be sealed off to prevent air from entering. The fermentation process is complete after one or two weeks. The maturing process of the drink, during which the flavour and taste intensifies, can take about six months.

The drink is served and consumed in a unique way. The host pours the fermented millet into a special vessel (the Tongba), adds boiled water (up to the brim) and, five minutes later, the drink is ready for consumption. The warm water, which now contains alcohol, is sucked out with a thin bamboo straw and when the Tongba becomes dry the host adds more hot water. The process is repeated until the alcohol is exhausted. This will take time!

Today, there are not only some family parties on my schedule but also a party with Kurt, a dentist from Hawaii, who did some trekking with Temba two years ago. He has returned to this area for a new project and this event needs to be celebrated. The striking effect of hot alcoholic drinks is commonly known, and soon I feel it spreading to every single cell of my body – in particular, the cells of my brain. However, I do quite well, and at the end of the party, I am the only one who still manages to find the way back to Sangbu's house...

DAY 21
CHEPUWA – HONGON
More family

We leave Chepuwa and follow a good trail through hilly terrain with lush, green pastures and seemingly hundreds of terraced fields where the local people grow millet, corn, some vegetables, wheat and barley. After a short and pleasant walk, we reach Hongon, where we will stay with Temba's mother-in-law for the night and employ Pimba, her youngest son, as an additional porter.

While Pimba lives in Hongon with his mother, helping her in the fields and with the animals, his older brother Lakpa (not to be confused with our cook), whom I have known for a couple of years, has climbed the professional ladder over the years. He started working as porter, became a kitchen helper and advanced to a cook position. Later, he became a trekking guide and finally a high altitude guide for expeditions. By now, he has guided several clients to the top of Mount Everest and regularly gets jobs because of his experience and knowledge. A fantastic career for a farmer boy from a faraway village in the Makalu region!

Pimba, Lakpa and Mingma, Temba's wife – they are a family with only three children and I wonder about this because families with five or more children are the norm in rural areas. And then there is the difference in age? Pimba is twenty-three years old and Lakpa is forty. Many women in Nepal give birth every second year at least. Temba tells me that his mother-in-law had been pregnant fifteen times, but that only three children survived. I can hardly believe what I hear, and I can hardly believe that this woman, who radiates so much joy, calmness, warmth and cordiality, experienced such dreadful strokes of fate.

Word of our arrival spreads rapidly through the village and locals come out to greet us. Soon, we are invited into their homes for countless cups of Tibetan Tea, Chang and Tongba. It appears that everybody is related, and it transpires that all local marriages occur between people from just two local villages: Chepuwa or Hongon. 'This is my cousin and his older sister, and that is my brother's mother-in-law with her granddaughter' Temba explains. It does not take long, but I totally lose track of who is who. While sipping my drinks, I cannot help thinking of the consequences of intermarriages. For related parents, there is a much higher risk of having children with health problems or genetic disorders than for unrelated parents, due to a lack of variation in their genes.

According to Jyllands-Posten, (27.2.2009), the risk of stillbirth doubles when parents are related. A study also analysed the risk of perinatal mortality (the child dies shortly after birth), infant death (the child dies during the first year), serious birth defects and severe or even fatal genetic disorders. On previous journeys to the Himalaya, I noticed that a particular genetic defect would often outnumber other disorders in a given area. In his book *The Kangchenjunga Adventure*, Frank Smythe wrote about Khangpachen, a small village in the Kangchenjunga area that, 'Among so small a population, the evils of intermarriage soon manifest themselves, and at the present time a number of its inhabitants are cretins… dwarf-like'.

Westerners often forget that endemic genetic disorders were once common in the Alps, too, described by ancient Roman writers. Later, the first Alpine mountain climbers and travellers recounted stories about coming upon entire villages of cretins.

DAY 22
HONGON – HATIYA
Farewell ceremony

The stage over the three passes Sherpani Col (6,180 metres), West Col (6,190 metres) and Amphu Labsta (5,845 metres), is considered to be difficult and dangerous by the local people. None of them would go there just for the fun of it. For them, it is a serious undertaking, entailing the risk of losing one's life in the snow and ice. In their Buddhist way of thinking, the consequences are severe. Some time ago, Pimba's brother Lakpa explained, 'We do not want to die in the snow. We may return as pretas'. Pretas, also called Yidak in Tibetan, are ghost-like creatures that are believed to have been jealous or greedy people in a previous life. As a result, they experience an insatiable hunger more than mere human suffering. Some Pretas can eat a little but never find food or drink, whereas others find food and drink but cannot swallow it. There are Pretas who experience the food to burst into flames as they swallow it, whereas others have to witness something edible or drinkable withering or drying up before their eyes, leaving them hungry all the time. This fact explains why Preta is often translated into English as 'hungry ghost'. In addition to hunger, Pretas suffer from undue heat and cold. In the summer, even the moon scorches them, and in the winter, the sun freezes them.

In Buddhism, Pretas represent one of six possible states of rebirth, and being reborn as a Preta is probably the worst thing Buddhists can imagine. The endless suffering resembles Christian hell, and although some Pretas can be nuisances to mortals, they are pitied by most people. Thus, in some Buddhist monasteries, monks leave offerings of food, money, or flowers to them before meals.

Despite the chance of getting some food or flowers from Buddhist monks, nobody wants to return as a Preta. This also applies to my group

Our farewell puja: Pimba's aunt, me, Temba, Lakpa, Pimba und Sonam *(from the left)*

– including me. Thus, in order to avoid a miserable existence in the life to come, the elderly women in Hongon perform a puja. We, and our journey, receive their blessings, chang is served, the gods receive offerings, and we get khatas.

A khata is a traditional ceremonial scarf, usually made of silk, with auspicious Buddhist symbols and mantras woven or inscribed into the fabric. In Tibet and Nepal, most khatas are white though one can find yellow ones as well. They symbolize purity, compassion and the pure heart of the person presenting them. These scarves play a vital role in ceremonies such as weddings, funerals, births, graduations, and arrivals and departures of guests. The offering of a khata is more than a simple gesture because it has its own protocol governed by tradition.

I am emotionally deeply affected by the ceremony and have to fight back a tear. Before setting out from Hongon, I attach the khata, that Pimba's mother presented to me, to my rucksack. It will join me all the way to Hilsa where I will leave it behind at the bridge leading to Tibet.

DAY 23
HATIYA – PIDING KHARKA
Leeches attack!

In Hatiya we employ Kinsang, who has crossed the three high passes leading into Solu-Khumbu before and is, in addition, familiar with the jungle-like forests we have to walk through. Since he, like everyone else, wishes to avoid being reborn as a Preta, the locals arrange another farewell puja.

The necessary shopping is time-consuming as it entails haggling and arguing over prices, and noon has already arrived when we finally leave the village. Rucksacks and dokos are filled to the brim and we are carrying an extra twenty-five kilograms of rice, twenty litres of kerosene, fresh vegetables, sugar, salt and tea. This is, hopefully, enough for the next ten days or so.

We have not been walking for more than one hour when it starts raining. Or, more accurately, pouring. Soon, the trail turns into a muddy stream and after squelching through the dirt for some time, we agree on pitching tents at the next kharka (shepherd's hut). Luckily, the next makeshift shelter is less than fifteen minutes away but, even so, we are soaked by the time we arrive. The smoke, leaking through the roof, tells us that people live here, and as soon as the tents are up we pay the shepherd's family a visit. For me, the visit is over half a minute later, the smoke inside the hut suffocates me and makes my eyes burn and so, while my team are chatting away, I get established in my little, red 'home'. It is then that I see my first leech on the journey.

From times immemorial, medicine has used leeches to remove blood from patients; the practice of leeching was well-known to the people of Ancient India and Greece. Both in Europe and North America, this treatment continued well into the nineteenth century and is, even today, still practiced occasionally.

There are about 700 species of leech, although only ten per cent of these are blood suckers. The Nepalese leech is one of the ten per cent. They are not dangerous as such, but they are an annoying aspect of trekking in the countryside during the rainy season and, in my opinion, they are particularly treacherous. The Nepalese leech is as thin as a thread and easily creeps through small holes, like those for shoe laces. They are ambush predators who wait in trees, bushes and high grass until they can strike prey – such as unsuspecting and incautious tourists – with their proboscises in a spear-like fashion.

Books and travel guides describe the most shocking stories about leeches and the leech-infested mountains and hills in monsoon. I have been to Nepal several times in the summer and while I cannot verify any of these stories, here are some basic rules to avoid getting sucked empty:

- Never walk around with bare feet
- Protect your head with a hat, or hood
- Avoid standing close to water or on wet grass
- Keep an eye on your shoes to be able to remove 'attacking' leeches
- Do not use Mother Nature as a toilet; or, if you must, avoid squatting down in wet grass.

So, what to do when a leech is attached externally? There are many methods to remove a leech, however, the little monster gets seriously angry about most of these methods and will regurgitate its stomach contents into the wound. Frequently, the vomit carries disease and thus increases the risk of infections. The safest method, therefore, is to wait until the leech is satiated with blood because then it just lets go and falls off –although this may take some time!

Today, I am lucky. None of these ugly, treacherous devils gets me.

DAY 24
PIDING KHARKA – NARI KHARKA
To cry 'wolf'

Torrential rainfall hit us last night, and I try to imagine what it must be like in monsoon time. Stop! There it is again, the meaningless pondering about the future. Why do I waste time mulling over the monsoon arriving in June? We are in the middle of April, and the sky is blue!

A long ascent awaits us, but I do not mind struggling uphill since the likelihood of being bothered by leeches dwindles the higher one gets. In my opinion, this fact is worth every single drop of sweat running down my face.

The location of a campsite generally depends on the availability of drinking water. Occasionally, we have to walk further than we actually want to, or we have to pitch tents when we would prefer to walk on for one or two hours more. Today, the latter is the case. We reach Nari Kharka in the early afternoon.

What to do with a free afternoon in the middle of nowhere? I decide to take a bath first because the nearby river provides some excellent spots for a decent washing session. Later, we set out together to explore the area close to the camp. Close to a clearing, Pimba discovers 'jungle spinach' and, since I am always curious about herbs and plants, I am excited by his find. The 'jungle spinach' resembles ground elder, also called goutweed, and gardeners in Europe label it the 'worst' of the garden weeds as it spreads so rapidly under favourable growing conditions. I know that the tender leaves can be used as a spring leaf vegetable for salads or they can be prepared like spinach. For our outdoor kitchen, the plant is a welcome change, rich in vitamins.

While Lakpa and Sonam cook the dinner, Temba narrates a story his mother had told him when he was a child:

'There was once a shepherd boy who grew bored as he sat on the hillside watching the village sheep. To amuse himself he cried out: 'Wolf! Wolf! There's a wolf chasing the sheep!' The villagers came running up the hill to help the boy drive the wolf away. But when they arrived at the top, there was no wolf in sight. The boy laughed at the sight of their angry faces. 'You shouldn't cry "wolf", shepherd boy, when there is no wolf', the villagers said. Then they went grumbling back down the hill. Later, the boy sang out again, 'Wolf! Wolf! There's a wolf chasing the sheep!' And to his naughty delight, the villagers all came running up the hill again to help him scare the wolf away. But when the villagers saw there was no wolf, they said to the boy, 'Save it for when there is actually something wrong! Don't cry "wolf" when there is no wolf!' But the boy just grinned and watched them go grumbling down the hill once more. Later he saw a real wolf prowling about his flock. Alarmed he leaped to his feet and called out as loudly as he could, 'Wolf! Wolf!' But the villagers thought he was trying to fool them again and so they did not come. The boy and the sheep were eaten by the wolf'.

Here, I am sitting in the Himalayan wilderness, listening to a story I had read at school. How has Aesop's Fable arrived in the Makalu area, I wonder. Was it brought here by missionaries?

The Christian church in Nepal is one of the fastest growing Christian communities in the world. Some time ago Temba told me that many people in his home village have converted to Christianity, and also one of his brothers has become a 'believer'. Does this mean that Buddhists are without belief? I do not think so, but some Christian groups describe them this way.

On a web site owned by a Christian organization *(http://joshuaproject. net/people-profile)* which leads research around the world and analyses the 'success rates' of conversions amongst different ethnic groups, one learns about the Loba, who 'live in the Kingdom of Mustang (Nepal) and remain one of the most isolated people in the world, both geographically and spiritually... Ask the Holy Spirit to cause the

Loba to become dissatisfied with their traditional religions, and to make them hungry for the Bread of Life'.

Now, it is not my intention to belittle the achievements of Christian organizations and individuals in general, but there are some black sheep around who show neither respect nor tolerance for others and their creeds...

DAY 25
NARI KHARKA – TOTRE
Nothing but forest

Today, while walking through the dense forest, I remember a book I had read a long time ago: *Natural History Handbook for the Wild Side of Everest: The Eastern Himalaya and Makalu-Barun Area.*

'The project area around Mount Makalu and the Barun river valley of northeastern Nepal represents an intact, but threatened, ecosystem in an area of globally significant biodiversity, the Eastern Himalaya. From tropical forests along the River Arun to the icy summits, the Makalu Barun National Park and Buffer Zone is the only protected area on earth with an elevation gain of 8,000 metres within thirty kilometres. As a result of this steep terrain, the Makalu/Barun area has twenty seven distinct forest types from almost every bio-climatic zone of the Eastern Himalaya.'

Through rivers and jungle-like forest

Because of the density of the forest, two men from the group walk ahead to cut a path through the vegetation with their kukris. The cutting power of this heavy, curved knife is astounding. It can be employed to cut firewood, open boxes, cut vegetables or hack down vegetation. Today it is in use for hours on end.

Now and then the jungle becomes less impenetrable and we make good progress for a while. However, we have to cross some rivers and, since nobody bothered building bridges, we spend a lot of time searching for fords or tree trunks that can be used as bridges. Unfortunately, I am, by no means, a professional high wire dancer – and certainly not with a rucksack on the back. Thus, I prefer to wade through the rivers whenever possible, and I consider myself lucky that they are neither deep nor particularly cold.

DAY 26
TOTRE – CAMP
Home of the Yeti

Since we left Hatiya we have not met one human being and, according to Temba, the chances of bumping into a local or even a tourist before reaching Yangla Kharka in four or five days are virtually nil. Our route runs far away from the main trail, and hardly anybody follows it. For Europeans, this may sound unreal since we are not used to roaming through a deserted wilderness. I derive great pleasure from being alone in the quietness of the forest. Well, the word 'alone' is not quite right because I travel with five other people, but walking behind them most of the time makes me feel alone.

The Makalu/Barun National Park and Conservation Area is the least populated and the least visited area of the High Himalaya. The ecosystem in this region is still intact and would make a perfect habitat for the Yeti. New evidence suggests that he lives here in the remote valleys that drop from the mountains surrounding Mount Makalu, but generally the scientific community regards the Yeti as a legend to be considered a form of parallel myth to the Bigfoot of North America.

Such creatures do, however, exist in the minds of many people – and in all cultures. In Afghanistan and Pakistan the locals fear the *Barmanou*; the *Almas* roams through Central Asia and in Scotland the *Fear Liath* makes the people shiver. The *Nuk-Luk* terrifies the Canadians and the *Hibagon* the Japanese, while the Norwegians have their *Trolls* and the Siberians their *Chuchunya* – just to name a few! No matter where all these creatures may exist, they have one thing in common, they are a metaphor for pristine, intact wilderness. As soon as the nature is being destroyed their habitat gets lost and, as a result, they will disappear. Thus, the world is richer if the possibility of the Yeti still exists and, to be honest, this is why I prefer a life with the Yeti.

Around lunchtime, we start a steep ascent on the north side of a mountain pass. The masses of snow at an altitude less than 3,000 metres surprise us, and with the top layer being frozen solid, we decide to use crampons for the first time on the journey. The climb is hard work, but I can easily cope with tough sections as long as I have a visible goal in front of me. Here, I can see the pass; a prominent saddle high above me. It comes as a relief that plodding through the ankle-deep mud of the jungle has come to an end.

DAY 27
CAMP – CAVE SHELTER
Magic?

The clear air of the early morning provides a stunning view of snow-capped peaks, and in the far distance, I can even make out Mount Makalu. But as ever, the weather deteriorates during the day and in the afternoon we are trapped in clouds and fog; the whiteness around us is virtually impenetrable. Under these conditions, we cannot walk on and, therefore, opt for a break. Full of optimism and confidence Temba declares that it will only take twenty minutes for the weather to change, and off he goes, together with Sonam and Lakpa. 'Good visibility in twenty minutes? That's rather unlikely', I mumble and find myself a flat, comfortable rock to sit on and relax. Less than five minutes have passed when they return with some plants and coax a fire to burn them. Soon, heavy smoke is all around us, and my group begins to recite mantras.

Incense burning is an integral part of all cultures and religions. The Germanic and Celtic people burned herbs during rituals, and frankincense is still used nowadays in Christian churches on festive occasions. In some Islamic countries, the faithful burn twigs from olive trees in mosques, and this method of purification is also known and used in Judaism. For Hindus and Buddhists, incense burning is a daily routine.

In Buddhism, this sacred offering is a way to honour the Triple Gem of the Buddha. The burning of incense results in fragrant smoke that

A cold morning with a stunning view

teaches us the necessity of burning off negative qualities within ourselves to reveal the pure self. The aroma of incense, typically derived from herbs, flowers and other natural sources, purifies the atmosphere and inspires us to develop a pure mind. The fragrances spread far and wide and this is compared with good deeds bringing benefits to many people. The rising incense dissolves into the air and, thus, subtly reminds the viewer of the transient nature of existence.

It turns out that also the clouds, surrounding us tightly a few minutes ago, have a transient nature of existence. They dissolve into the endless universe, and we have perfect visibility for the rest of the day; I can hardly believe it.

DAY 28
CAVE SHELTER – CAMP
Footprints in the snow

My day starts well. I wake in a cold and damp sleeping bag, spill noodle soup in my tent, honey drips from my chapatti onto my socks and I discover that the shampoo bottle leaks – the sticky liquid filling the side pocket of my rucksack.

In all honesty, I am not in a particularly good mood when we set out for the pass and not even the white peaks, rising into a stainless blue sky, can pull me out of my emotional void. My gloomy mood does not improve when I disappear in a hole up to the hips between two boulders. When reaching out for a branch it breaks (of course) and, unable to get out on my own, I cry for help. The worst, however, is yet to come. One of my walking poles takes on the shape of a V when leaning on it with my entire body weight. It is totally useless, and I do not know how I will cope with long walks downhill without it. When and where will I have an opportunity to purchase a new one? Today everything seems to go wrong; completely wrong. I am aware of the fact that this underlying anger nagging at me will only result in additional disasters. It is a classic example of so-called 'negative thinking'. But how do you get out of a negative mind-set when everything goes wrong?

On getting to the pass about one hour later, we look in vain for a trail leading down to the valley. After a while, however, Pimba detects footprints; very uncommon ones. 'Snow Leopard', he shouts excitedly. Snow Leopard? I can hardly believe this to be true, and the first thing coming to my mind is *The Snow Leopard* by Peter Matthiessen; a book worth reading.

Snow Leopards live in the mountains of Central Asia and in the Himalaya at an altitude between 1,250 and 6,000 metres, depending on the time of the year. Worldwide, the estimated number of these big cats is somewhere between 4,100 and 6,600, placing them on the list of endangered species. Worse still, the number of those that are able to reproduce is probably less than 2,500.

To be able to cope with the harsh conditions at high altitude, Snow Leopards have long, thick fur and small, rounded ears to minimise heat loss. The wide paws, covered by hair on the underside, are ideal for walking on snow and steep terrain, and the long, flexible tails help them to maintain balance. Although Snow Leopards are carnivores in general and capable of killing animals three to four times their size, they have never been reported to attack humans. As a matter of fact, they are among the least aggressive big cats and can be easily driven away from domestic livestock. Snow Leopards readily abandon their kills when threatened and may not even defend themselves when attacked.

Of course, it would be an unbelievable experience to see a Snow Leopard in its natural habitat, but these animals are extremely wary and so I am happy to admire the footprints in the snow, thinking about a big cat that, probably, crossed the pass less than ten minutes ago. It is an exciting thought, and suddenly, the minor misfortunes of the day are forgotten.

The state of positive thinking only prevails until dinner. One of the kerosene stoves no longer works because of a damaged valve and all attempts to repair it fail. The ten-day walk to Chhukung in Solo-Khumbu crosses two 6,000 metre passes. Having one intact stove only for an undertaking like that cannot be considered a good plan...

DAY 29
CAMP – YANGLA KHARKA
Attention, landslide!

Eventually, we get down to the bottom of the valley. I am deeply relieved because the trail beside the river looks easier and better than the one we descended this morning. Looking easier, however, turns out to be a deception. In several places, there are a couple of apparently new landslides and there is a sense of danger and peril about this section. We cannot see any footprints and wonder how safe it is to cross the slides. I peer longingly at the other side of the river where the main trail runs, but even on that path, several stone avalanches have swept down into the river, burying the trail.

I think of all the tourists who are reported missing in Nepal every year. Personally, I am convinced their disappearances are, in many cases, the result of accidents and not robbery or murder. Even during the civil war (1996-2006), when the Communist Party of Nepal (Maoists) fought against the monarchy and the Hindu caste system, not one single tourist was attacked or killed. However, at least 15,000 Nepalese people lost their lives during the decade-long conflict.

Before arriving at the hamlet of Yangla Kharka, we have to cross half a dozen massive landslides with hundreds of tons of rock and stone piled up on the steep hillside above us. These piles do not look particularly stable and we consider ourselves lucky that there has not been any rain in this area the last few days. Dry weather minimizes the possibilities for further landslides to sweep down, but it does not eliminate the risk completely. Knowing this, I cannot help peering anxiously at the rocky chaos higher up, as if staring would prevent a landslip rushing down and sweeping me off to a sudden death. In the rain, this section would be extremely dangerous. Finally, the valley widens, and the steep walls step back – the imminent threat is over.

The small settlement at Yangla Kharka lies in the very centre of a wide grassy basin and consists of a few scattered houses, a hotel and a shop where it is possible to stock up on basics. Most trekkers who are on their way to Makalu Base Camp take a break at Yangla Kharka, either to have lunch or to stay for the night. We hope to get a stove here, be it a new or a second-hand one. It turns out that we are lucky!

DAY 30
YANGLA KHARKA – LANGMALE KHARKA
Nepali women and mountaineering

It is a short but memorable day of trekking. Great rock walls tower above us as we continue our walk through a vast U-shaped valley. Snowy peaks are the stunning backdrop to Langmale Kharka where Pasang Wangsu Sherpa runs the small hotel, a converted shepherd's hut. He is seventy-three years old and his deeply wrinkled face resembles old leather, worn and torn by wind and sun. Over a cup of tea, he proudly tells me about his ten children: three sons and seven daughters. Of course, I am curious about the number of his grandchildren. 'A hundred, maybe?' he replies with a boyish grin revealing two lonely, brown front teeth.

Pasang, however, is not proud of his family alone, but also shows pride when talking about his 200 yaks and 300 sheep which give him the status of a wealthy man. He radiates happiness and seems to enjoy his solitary life in the mountains. In one way or another, he reminds me of a Swiss mountain guide who once termed the mountains his church and his religion, his joy and his home. I believe that this applies to Pasang too. Some of his children live abroad, and they invite their father to England, France and even America, but he does not want to leave his home in the wilderness. Later he tells me that his wife does not share his genuine enthusiasm for this remote place; she lives further down the valley in a village.

In the afternoon, he invites me into his private room where the walls are decorated with newspaper articles and photos depicting a mountaineer on top of Mount Everest. In every picture, the climber holds a photo of our host high up. Pasang explains, 'This is my daughter Lhakpa Sherpa. She has reached the summit six times and is on her seventh expedition right now. Lhakpa always takes a picture of me to the top'. His daughter? I am surprised.

In 2000, Lhakpa followed in the footsteps of many other women who had climbed Mount Everest, but only one Nepali woman had reached the summit before, Pasang Lhamu, who later died on her way down. Today Pasang Lhamu is honoured as one of fifteen national heroes of Nepal, along with past kings and religious leaders. To honour her achievement, a mountain has been named after her – Jasamba Himal (7,315 metres) has been renamed and is now known as Pasang Lhamu Peak. Even a road was named in her honour, the Pasang Lhamu Highway connects Trishuli and Dunche.

Pasang Lhamu became a symbol of hope and a metaphor for the realisation of dreams. Any Sherpa woman who climbs now sees her as the trailblazer. She became the inspiring heroine for Lhakpa, our host's daughter; a heroine Lhakpa longed to emulate. Pasang had to fight against prejudices from all sides. As a teenager, she often joined her father, a mountain guide, on expeditions as a kitchen helper, as a girl with a secret dream, to reach the summit of Everest. Her family considered her wish reprehensible and presumptuous.

By the time she was thirty two years old, Pasang Lhamu had three children and a husband, and she had attempted to climb Everest three times without success. However, she was not a person to give up that easily. She summited Everest on the 23rd of April 1993.

Lhakpa's father does not seem to have any problems with the fact that his daughter has turned her back on the traditional role Nepalese women are expected to accept. He is proud of Lhakpa, who has dedicated her life to the mountains and the sport of mountaineering. Yet, she is not the only one in the family who longs for the high peaks. A couple of years ago, she reached the summit together with her sister Mingkipa and her brother Mingma*.

DAY 31
LANGMALE KHARKA – MAKALU BC
Living here forever?

Waking up in the morning to the soothing sounds of Buddhist prayer songs and flapping prayer flags is one of the things I like best when travelling through the Himalaya. Even before my alarm goes off at six o'clock, I hear Pasang reciting mantras, and the smell of burning juniper reaches my tent. In Himalayan areas inhabited by Buddhists, one can experience this traditional incense burning ritual everywhere. In towns and villages, the clouds of smoke wafting through the streets and lanes can cause breathing problems, but here in the fresh and crisp air of the

---

* In Nepal, names do not tell the gender of a person. Frequently the name is related to the day of birth. Some examples: Mingma – Tuesday, Lakpa – Wednesday, Pasang – Friday, Pemba – Saturday.

mountains, the wind carries the clouds away quickly. The peaceful atmosphere instils a sense of safety and mental calmness I sometimes miss in my busy life back home.

Before my tea arrives, I roll out of my sleeping bag, grab my camera, crawl out of the tent and stroll up a boulder-strewn slope behind the hotel. The valley we follow up to Makalu Base Camp is often referred to as the Yosemite of the Himalaya and it is truly an impressive and exciting place to be. The view towards the summits of Peak 3, Peak 5, Peak 6 and Chamlung, dominating this unique panorama, is breathtaking. They gleam in the bright light of the morning sun and their beauty helps me to understand why Pasang neither feels any compulsion nor want to give up his secluded life at Langmale Kharka. What would happen if I decided to stay here? I could enjoy a life close to nature, marvelling at the beauty of the mountains and selling tea and cookies to the few tourists coming up from Yangla Kharka. The old dream of getting away from the hustle and bustle of civilization and leading an austere life instead comes up; again. As a matter of fact, this idea has haunted me for decades. After all, what do I really need? For thirty days, I have been travelling light: rain clothes, one down jacket, one woollen jacket, three T-shirts, two pairs of trousers, four pairs of socks and some underwear; this is enough. When thinking of my closet, which is virtually bursting at the seams, I cannot help laughing about myself. Why do I have all this rubbish? To be quite honest, I do not know.

On the way to Makalu Base Camp

For a fleeting moment, I tinker with all sorts of ideas, but in the end, hot showers, washing machines, hair dryers and all the other electronic gadgets that make life easy in the West turn up before my inner eye and I know that I will never be happy here for long. For a relaxing holiday, however, I cannot think of anything better than strolling through the Himalaya, living in a tent and eating Dhal Bhat twice a day.

On the way to Makalu Base Camp, we walk through the most beautiful landscape of the Makalu Barun National Park with the snowy peaks lining both sides of the valley bearing a resemblance to the trees along a boulevard – trees guiding a traveller to his or her destination. Finally, Peak 3 and the snout of the West Barun Glacier round off the scenery, and for a fleeting moment, a sense of infinite happiness sets in.

DAY 32
MAKALU BC
Rest day

We plan to have at least one rest day at Makalu Base Camp. Even after having been above 5,000 metres a few times before, we know it is vital to be fully acclimatized before setting out for the three high passes. It is also necessary to sort out all the climbing equipment, do some basic training with knots and karabiners for Pimba and Sonam, who had never done any climbing before, and to eat and eat and drink… a lot.

M.B.C. is not the place one would choose for relaxation – even given that it is one of the most spectacular viewpoints in Nepal, with the massive bulk of Makalu rising to a perfect pyramid summit. It is a dreary and desert-like plain, without any trace of green at this time of the year – just sand, stones and rocks. The fine, white sand is the result of the grinding work performed by the glacier and the nearby river over thousands or even millions of years. The wind blows incessantly. In the morning, it blows from the north, where Mount Makalu dominates the landscape. Around noon, it shifts direction and blows from the south, covering everything with dust.

Only two of the hotels are open and, needless to say, my team opts for the one that is run by a charming seventeen-year old beauty. Normally, she attends a boarding school in Kathmandu, but during her holidays, her parents, living further down in the valley, send her up for a couple of weeks to work here. Her laughing and giggling fills the place, and there is a lot of flirting with my team members and with the two Nepali

Our charming hostess

mountain guides who are here assisting members of a Chinese Makalu Expedition. The guides have come down from High Camp to rest for a few days before the final push for the summit.

I wonder how many seventeen-year old European girls could cope with life in a remote place like this one; a five-day walk away from the next road. It is a life without internet, Facebook, mobile, electricity, running water and heating. After sunset, the room temperatures drop to somewhere around zero in the kitchen where she sleeps.

DAY 33
MAKALU BC – CAMP
Falling rocks I

Sherpani Col Base Camp is only a two-day walk away from Makalu Base Camp, and knowing that they will be short days, we feel no need to hurry. While Temba, Pimba, Lakpa and Sonam are eating breakfast in the main part of the hotel, I take numerous pictures of Mount Makalu, its white flanks gleaming in the sun. All of a sudden, a deafening noise interrupts the serene calmness of the early morning. It sounds like an avalanche – a rock avalanche – right behind me. Quickly, I turn round and witness the disaster at full length. Less than five metres away from me, two walls of the annexe are collapsing.

Collapsed walls

Many houses in Nepal are built of unhewn stones and, due to poverty or unavailability, cement and mortar are rarely used. I have always wondered about their stability. This very moment, I get an answer to my question. The walls are less stable than I thought.

A blessing in disguise. None of us were in the annexe when the walls collapsed and stones buried the sleeping places. We excavate the equipment and inspect every single item we secure. Deep relief sets in when finding everything undamaged apart from a washing bowl.

When leaving M.B.C. in bright sunshine, we are in a good, cheerful mood. In the beginning, the trail runs across Alpine meadows at an easy gradient, but soon the path gets lost between thousands of massive boulders and it becomes difficult to settle into a steady rhythm. I experience shortness of breath when climbing over the rocks, or when trying to leap from one to the next. The boulder field seems endless and, though we walk for less than four hours, I am tired when arriving at the next campsite. As soon as Sonam and Pimba have set up my tent, I crawl inside, switch on my iPod and relax to some music. Half an hour later, when Pimba comes with a cup of coffee, my tiredness has gone.

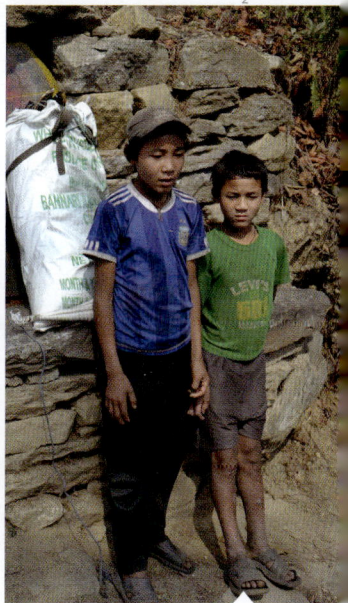

1 My cousin Reinhold and me, on a mountain trip above Garmisch-Partenkirchen, Germany in 1964. 2 Children in a class run by *Autism Care Nepal* – the organisation founded and run by parents of children with autism. 3 Local children from well-off families playing Caramboard. Lapka and Nabin watched from a distance with neither a smile nor laughter.

4 Lakpa (13) and Nabin (11) Sherpa, working as porters and regularly carrying thirty-kilogram bags of rice up and down the hills. In Nepal it is illegal to work under the age of 15.

1 The village of Ghunsa is one of the main villages in the Kangchenjunga Conservation Area.   2 Our ice-cold hotel in Lhonak.
3 On the way to Kangchenjunga Base Camp.   4 Lali Gurans.

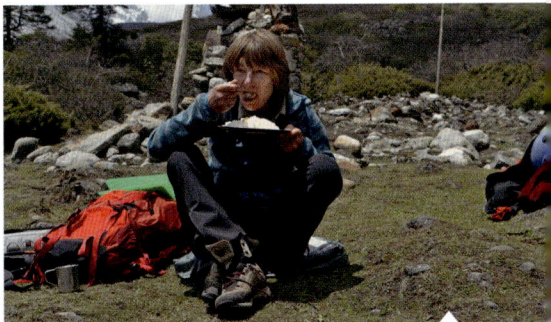

1 Temba's mother-in-law. 2 Everybody wants to see my pictures! 3 Pimba is snow-blind.
4 A lunch break with my beloved Dhal Bhat! 5 Pasang Wangsu Sherpa, 73-years-old and radiating happiness.
6 The 'shopping centre' in Ghunsa. 7 Yartsa Gunbu – 'Nepalese Gold'.

1 Terrace café in Ding Boche – where instant coffee tastes like heaven…   2 Climbing up Amphu Labsta – at 5,800 metres, every move was a 'big thing'!   3 Tengboche monastery.   4 Early start from Thyangbo Kharka in bright sunshine.

1 Bigu Gompa.

Sherpani Col Base Camp

DAY 34
CAMP – SHERPANI COL BC
Not the only one

Over the last few days we have repeatedly heard rumours about other trekkers walking the Great Himalaya Trail. Of course, I know about an organized group from Australia crossing Nepal. They had started in Taplejung about four weeks before I set off and like me, they are on the way to Hilsa; a small hamlet on the Tibetan border. The local people, however, do not refer to the Australians walking ahead of me. The rumour goes that there is someone behind us! Strangely enough, this affects me. I am not known for being a competitive person, but I liked the idea of being the only lunatic on the Great Himalaya Trail. By the look of it, I am not, and I begin to speculate on who the other person is.

When ascending to Sherpani Col Base Camp, we have a marvellous view down into the valley and of the isolated summit of Mount Makalu. The peak resembles a four-sided pyramid. The panorama is stunning and I take frequent breaks to feast my eyes on the landscape, to enjoy the quietness of the mountains and to wonder about the perfection of nature. During one of these rests, I make out four fast moving specks further down the slope, and when they come closer I am able to judge from their clothes that two of them are tourists. Obviously, we are not the only ones, at least not the only ones to cross Sherpani Col tomorrow.

On arriving at the Base Camp – where my altimeter shows 5,688 metres – the sun is about to disappear behind a ridge and after a few minutes it feels as cold as a freezer. Strong gusts of icy wind tear at the tents when we try to set them up, and when they are finally pitched we fear they will be blown away. I climb into my tiny tent, snuggle into my thick down sleeping bag and listen to the whining wind tearing and tugging at the fabric. Would I leave this comfortable place? Not for love nor money! But I do so for a Dhal Bhat.

The other party has arrived at the Base Camp too. It is a couple from Switzerland, who are following the Great Himalaya Trail, and their two porters. Nicolas has planned to walk many parts of the route on his own. I am deeply impressed and admire his courage and determination, but there is no jealousy. He will have to carry a heavy rucksack with probably thirty kilograms or more. I remember trips with an overloaded rucksack strapped to my shoulders, trips carrying a tent, cooking gear, equipment and food for ten days… This was not always great fun! Luckily, employing a few porters to help me is a legitimate and wise decision at my age; I feel privileged.

DAY 35
SHERPANI COL BC – WEST COL BC
A plan fails

Today's aim is to cross both Sherpani Col (6,180 metres) and West Col (6,190 metres). The plain between the two passes lies at an altitude of over 6,000 metres, and is not a particularly good place to camp – we could easily be trapped there by bad weather for several days, which would increase the probability of suffering from altitude sickness. We decide to utilise the perfect weather and cross both passes in one big push and so yesterday resolved to make an early start.

At five o'clock, my 'good morning' tea arrives and, less than twenty minutes later, we eat enormous servings of Dhal Bhat in the kitchen tent. There will hardly be time for prolonged breaks.

At high altitude it is important for me to find a suitable rhythm and to follow it. Usually, I walk slowly to avoid fatigue and take short breaks regularly to eat and drink. After a long plod up a snow-covered slope, I reach the rocky area below Sherpani Col, and half an hour later the view

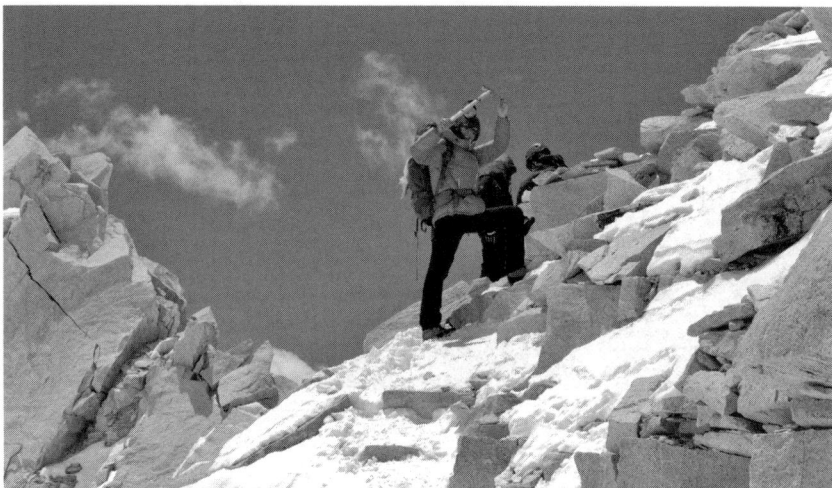

On Sherpani Col

down to the Barun Glacier unfolds in front of me. I am surprised to have enough energy left to take numerous pictures and to smile happily and proudly into the camera lens when Temba takes some photos.

The Swiss couple who had left Base Camp after us has also arrived on the pass, and as a result, the sharply edged ridge is pretty crowded, with ten people standing or sitting on three square metres at the top. Stumbling over a rope or a rucksack would end in a disastrous fall; probably a fatal one. Together, we establish a belay, fasten the rope to it and begin the abseil. It takes quite a while for everybody to reach the glacier and for the loads to be lowered down. What follows is a pleasant stroll in the sunshine over to West Col, on flat ground. Looking at the perfect symmetry of Baruntse (7,100 metres) dominating the mountain scenery to the right makes me want to climb the mountain. I try to imagine what it would be like to gaze down into the maze of Himalayan valleys from the highest point in the area; without being surrounded by higher peaks. Well, that will have to wait until my next life... probably.

Arriving at West Col, however, we have to face reality. A Japanese expedition, climbing Baruntse, is about to leave, and at least twenty porters are busy hauling down tons of equipment they no longer need. They use the only belay point, and we wonder how long will it take until they are finished? Do we have to spend a night here at 6,100 metres? I do not like the idea, but my feelings cannot prevent the inevitable.

Next day we learn that they had been lowering down people and gear until midnight.

Camp below West Col

DAY 36
WEST COL – HONKU BASIN
Snow blindness

It was one of the coldest nights in a tent I can remember. My thermometer stopped working, the sun cream froze in the plastic bottle and when moving in the tent, the frozen condensation rains down on me. Such a 'high altitude morning shower' awakens me quickly. However, as I pull the frozen flap of the tent aside, all discomfort is forgotten. A cloudless, blue sky promises a perfect day for the crossing of West Col.

I am just about to leave the tent, when Pimba comes and tries to tell me something but the only word I understand is 'eye', and I have to ask him to get Temba to translate; Pimba is snow-blind. This news comes as a shock, not only for me, but for all of us. Yesterday, we had told him again and again never to take off his sunglasses when outside, but he had ignored our warnings. Even though it is 'only' temporary eye damage, caused by snow reflecting UV light, it is terrible to be snow-blind, as I know only too well.

At the age of eighteen, I went skiing in the Alps without sunglasses. In the evening, I watched TV but soon wondered about the poor quality of the screen; all the pictures were blurred. Maybe time for bed, I thought and looked at the clock hanging on the wall, but I could no longer make out the numbers. It was then that I realised that there was something wrong with my eyes – and so decided to go to bed.

Of course, worrying thoughts about my impaired vision kept me awake, and so I felt the excruciating pain develop. It felt as if sandpaper was scratching over my open eyes. I got up and switched on the light, but there was no light any longer; I was blind. It took about one week with medical treatment to get better, or at least well enough to find my way to school – where I had to swap places with a girl sitting right in front of the blackboard.

We have no idea how serious Pimba's snow-blindness is but here, just below West Col, we have no way to get medical help for him. The only thing we can do is to redistribute the five loads so four people can carry and freight them up the steep rocky area to West Col, which is no more than 150 metres away from our camp. The last few metres are dangerous because a slip would result in a fall. Temba does an excellent job when leading Pimba safely to a place where he can sit down and wait until everything is ready for the abseil. It is clear that Pimba is scared, and we understand his fear. From West Col, we have to get him down a steep, almost 160-metre-high wall of ice and rocks – without him being able to see anything. His lack of experience makes the situation even more difficult than it already is. All he knows about abseiling is based on the instructions we have given him on the walk up and from the short abseil we made on Sherpani Col. I can only give him my helmet to make him feel safe.

Luckily, we have enough rope to install two fixed ropes and, with Temba by his side, they start the long abseil, stopping again and again as Temba encourages Pimba, calming him down when panic sets in after a slip on the icy surface. It takes them ages to get down to the bottom, and all of us breathe a sigh of relief when we see them waving at us from far below.

The glacier covering the Honku Basin poses no real risks because the few crevasses are clearly visible, and so we do pack the ropes away for the walk to our next camp. Though Temba has added some of Pimba's load on top of his rucksack, Lakpa and Sonam have to carry more than usual and are exhausted on arrival at the camp. Yet, as soon as the tent is erected and water for the tea is boiling in a pot, I can hear joyous laughter, songs and countless jokes once more. Where do they get their strength from? I wonder.

DAY 37
HONKU BASIN – AMPHU LABSTA BC
Alternative route?

Of course, given Pimba's condition, we had discussed the topic of alternative routes yesterday, but there is no real alternative as such.

It would be possible to proceed from Honku Basin through less demanding terrain to Lukla, where medical help is available, but it would take at least one week to reach the village. The other possibility is to follow our planned route over Amphu Labsta, although crossing the pass is technically more difficult than the two passes we have already navigated. It is a dilemma.

Statistics say that ninety per cent of the things we worry about are based on mere assumption, without any real cause for worry. Furthermore, nine out of the remaining ten per cent will never occur. Luckily, this proves to be correct. Early in the morning, Pimba comes over to my tent and smiling brightly, declares, with Temba's help, 'I can see a bit better. The pass we have to cross tomorrow will be no problem'. This is an enormous relief for us. We leave our camp on our planned route to Amphu Labsta Base Camp.

Countless steep moraine hills have to be climbed, and a couple of times I wonder why there are two cameras and a laptop in my rucksack in addition to the necessary gear. Wouldn't it be enough to enjoy happy moments and take them home in my heart? Photographs are two-dimensional, without smells, without sounds... no cold wind makes the spectator shiver, no sun makes him (or her) sweat and nobody gets wet when rain pours down... pictures lack the overall 'feel' of the moment; they are no more than sentimental reminiscences.

Since breakfast I have suffered from stomach cramps and have little energy for philosophical ponderings about the intrinsic value of pictures. I feel weak and tired, and the pain makes my life miserable while plodding behind the others. The situation reminds me of the day when Temba and I had walked to Kangchenjunga Base Camp. As then, my body seems to cry out 'NO' whenever I take a sip from the bottle. Of course, our drinking water gets boiled, but this is useless without a pressure cooker. The higher one gets, the lower the air pressure becomes, resulting in a lower boiling point (0.5 degrees per 100 metres difference in altitude). At 5,000 metres, joyously dancing bubbles in your cooking pot do not indicate that the water is boiling properly. As a thermometer will reveal, it will only have reached seventy-five degrees – the water is not hot enough to kill germs. I tried to explain this fact to Lakpa on a couple of occasions, but I am afraid that he did not get the point. Maybe I should try again using different words.

THREE

SOLU-KHUMBU

Mount Everest seen from Tengboche

DAY 38
AMPHU LABSTA BC – CHUKHUNG
Through the ice

We are not the only ones to cross the Amphu Labsta pass today. A British group, who have climbed Mera Peak, are also on the way to Chhukung. When seeing them leave the camp as early as six o'clock, we dawdle around for some time. According to our information, there is only one good belay point for abseiling from the pass, and it does not make sense to me to queue up like mountaineers climbing Mount Everest. When we finally leave the camp, the British people are already out of sight.

Soon we reach the snow line, where I try to figure out the route ahead. All I can see is a barrier of ice in front of us. 'Is there a chance to avoid this bit?' I ask Temba. 'No. Sorry, we have to climb up here', he replies. I am not the most experienced ice climber and thus the following sixty minutes are full of suspense.

Technically, this ice section does not pose any real problems, but combined with an altitude of almost 5,800 metres every single move, that normally (and by 'normally' I mean at sea level) would be reasonably easy becomes a 'big thing'. All my muscles scream for oxygen, but up here there is only fifty per cent as much of that as I would like and I feel its lack whenever I have to dig my ice-axe into the hard ice, or whenever I have to make a big move… My heart beats at the speed of a machine gun. Sometimes it takes up to one minute to gather my breath; to recover the energy to move on. How do mountaineers cope at 8,000 metres? At that height, the air only contains thirty per cent oxygen.

The gradient of the snow slope eases below the pass and we pick up the pace. On arriving at the highest point, we notice that the British group has left behind their snow stakes, or 'pickets'. They are widely used in the Himalaya but their holding power is generally considered somewhat dubious. But we only have a couple of snow stakes and using theirs saves us some time, ensuring we will make it to Chhukung today. Back to civilization.

The long shadows of the afternoon sun accompany me as I arrive at the first houses of the village; real houses. The signs of the hotels carry promising names and it is difficult to decide which one to choose. Finally, I move into one of them and hope that my team finds me there, sitting in front of the warm fireplace.

To my disappointment there is no network coverage in Chukhung, but from locals I learn that one can make a phone call from the other side of the valley entailing a 200 metre climb up a hill. I have been walking for twelve hours today and tiredness has spread to every single cell of my body. Nevertheless, I set out for this last climb to call my parents who are deeply relieved to hear my voice from the other side of the world after forty days. Afterwards I phone David, my boyfriend at that time, and his shouts and congratulations are loud enough to have reached me without a phone. He seems to be happy and proud that we made it safely over the passes.

Total darkness has settled in by the time I stagger down the slope and stumble back to the hotel. Now, I am totally done… for today.

DAY 39
CHHUKHUNG – TENGBOCHE
An advert: 'XXX gives you wings'

It is a strange feeling to wake up in a proper room after sleeping in a tent for several weeks. I listen to the creaking of the wooden floor when people pass my room. Padlocks squeak, tables and chairs scrape and there is music from a radio. For a quarter of an hour I just lie in bed and try to get used to civilisation, until hunger forces me out of my sleeping bag and I walk to the dining room where breakfast is waiting. Here, the rumour goes that there are cafés in Dingboche, the next village, serving cappuccino! Nobody can stop me now. Friends know about my addiction to cappuccino, and the thought of strong coffee, topped with delicious white froth, makes me rush off only five minutes later.

'XXX gives you wings', an advertisement promises. I cannot confirm this statement, since I have never tasted XXX, but if it is true, people probably feel like me when I 'fly down' to Dingboche. I even manage to pass several locals! That should tell you a lot.

On arriving in the village, I hurry into the first terrace café, take a seat in a comfortable, blue plastic armchair and order my favourite drink: cappuccino. The fact that the owner serves instant cappuccino would disappoint me back home, but here, it does not. I am in heaven. I sip, slowly… slowly…

The enormous number of tourists passing the terrace soon brings me back to earth after my visit to culinary heaven. Over the course of one hour in the café I see more trekkers than on my thirty nine-day journey from Taplejung to Chukhung. This does not come as a surprise because the Solu-Khumbu region is the top destination for trekking tourists visiting Nepal, and every year hundreds of thousands stroll up and down the trails.

The walk down to Tengboche turns into a coffee drinking binge. I order a cup of coffee here, drink some café latte there and ask for more cappuccino whenever I round a bend and a cosy hotel pops up in front of me. My will is not strong enough to resist the temptation of sitting down in cafés along the trail. What a change! In 1990, when I was here the last time, there were no terrace cafés with mountain views lining the track, and luck was needed to find a lodge where the owner served instant coffee – which was usually then too weak and too sweet.

Since Lakpa has a relative working at a hotel in Debuche, situated below Tengboche, the group wants to stay there for the night. I stuff my sleeping bag into the rucksack and walk up the last hill to the famous Buddhist monastery of Tengboche. The place offers one of the most stunning views in this area, and I hope for a chance to take a couple of good pictures in the early morning. While climbing up to the ridge, I remember the run-down lodge that was there twenty-five years ago. What does the place look like today? In those days, the rooms were infested with fleas and bedbugs, the kitchen was full of smoke and the food lousy. But even then, one could take a shower – albeit an outside one. Some plastic sheets were attached to four wooden poles, and when ordering a hot shower, a monk would come with hot water and pour it into a large bucket resting on a makeshift scaffolding. One only had to unplug the hose dangling down from the bucket, and the 'shower adventure' started. In summer, this was pretty good fun, but otherwise one had to be tough as the strong, cold wind found its way from all sides into the 'bathroom'.

Today, Tengboche greets me with new hotels, a 'German Bakery' with spotlessly clean rooms, solar showers and a cosy café. There are heated dining halls, extensive menus and steaming hot towels with which to clean one's hands before eating. The toilets are clean and there is electricity and a mobile tower!

DAY 40
TENGBOCHE – NAMCHE BAZAAR
'If I had a hammer...' Johnny Cash

I get up early in the hope of having a chance to take some good pictures of the mountains and the monastery in the clear air of the morning. But reality and wishful thinking do not always go together and Mount Everest is hardly more than a blurred, bulky something in the far distance. However, Mount Everest is Mount Everest, and I take many pictures – only to delete them later.

On my way to Namche I turn around frequently, still hoping for a better view. The weather situation, however, stays the same and I finally concentrate more on the restaurants and cafés along the trail.

Way before reaching the first houses of Namche, I hear the sound of hundreds of hammers; or at least that's what it sounds like. 'Cling, cling, clang... cling, cling, clang...' I am reminded of Johnny Cash's song: 'If I had a hammer...'

Namche has been an enormous construction site for a long time, and over the years people have built new hotels, shops, restaurants and private homes. The stone for this comes from a quarry above the town where it is hacked into a rough rectangular shape. Afterwards, a seemingly endless stream of porters carry the stones to the centre of Namche where craftsmen hammer them into the right size and shape. 'Cling, cling, clang...' Still, there are no machines.

'Cling, cling, clang...' Everywhere men sit under differently-coloured tarpaulins protecting them from the glaring sun and the wind. They work with hammers and chisels and I wonder how long it takes to produce one

Worker in Namche Bazaar

perfectly-shaped stone for a building. How much does a craftsman earn for a job that probably leaves him deaf after a while or half blind? I see only a few workers wearing glasses to protect their eyes against flying stone chips and none of them are equipped with hearing protection.

Finally, I reach the hotel, and after having moved into my room, I brazenly take advantage of modernity that has arrived here and I treat myself to a long, hot shower; the first one for forty days. Later, I put on my last clean set of clothes and join my team in the town centre to celebrate the safe crossing of Sherpani Col, West Col and Amphu Labsta.

DAY 41
NAMCHE BAZAAR
Rest day

I slept deeply after returning from the party and today am looking forward to exploring Namche after breakfast, curious to see how the small town has changed over the last two decades. But before leaving the hotel I have a quick glance in the bathroom mirror, which tells me that I have more pressing business – a haircut.

Strolling down the 'Main Street' reminds me of Thamel in the '80s. Pink Floyd and Creedence Clearwater Revival blare from the loudspeakers

Namche in 2012

in cafés and restaurants and the shops and stalls offer the typical array of souvenirs: woollen jumpers and socks, necklaces and bracelets, singing bowls and brass statues. In one of the narrow lanes, I discover a sign for the *Highest Beauty Parlour in the World* and decide to give it a try. I am surprised to see that the place is run by a male beautician, but he seems to know his business and half an hour later my scruffy hairstyle has been turned into a stylish one. 'What about a facial treatment?' He asks, and when I agree, he begins to apply creams, ointments, lotions and oils to my sunburnt skin. I hardly recognize myself in the mirror two hours later. The beautician did an excellent job.

Back at the hotel in the late afternoon, I have a long conversation with Anu, the owner of the hotel, about all the changes Namche has experienced over the last two decades. Together, we look at pictures, stored on my laptop, showing Namche in 1990, and Anu tells me that not many of the old houses in the centre have survived the construction boom. I also learn more about other things that have been going on since my last visit more than twenty years ago.

While talking about the 'good old times', which were not good after all, one of the other guests orders a yak steak. Anu smiles, and later he explains that some time ago the local Mothers' Group* which is powerful in this area, succeeded in banning the killing of animals altogether.

---

* Mothers' Groups exist all over Nepal. Usually, local women meet once a month to discuss changes, improvements and developments. There are several organizations offering lectures about, hygiene, family planning, nutrition, reading and writing to these groups.

Thus, all the yak steaks served at hotels in Namche are beef steaks. The owners of hotels make a joint order at a butchery in Kolkata from where the steaks are freighted, nicely packed and deep frozen, to Solu-Khumbu, where they miraculously turn into yak steaks.

One can argue if this is cheating tourists or not. I remember the local market where chunks of meat lay on the ground, flies buzzing round, dogs sniffing at them... If I liked meat, I would opt for the frozen beef from Kolkata.

DAY 42
NAMCHE – THAMO
Change in the Weather

After a sudden fall in temperature with snowfall last night, the town is covered under a thin layer of new snow this morning. The weather forecast predicts bad weather and more snow for the following seven days. This does not sound good. The approach from Namche to Tashi Labsta (5,760 metres) is considered both magnificent and demanding as it is not only high (offering stunning views), but also glaciated and prone to rockfall. Additionally, the glacial areas are constantly shifting, and the route can be different from one season to the next one. Thus, the pass is not to be taken readily, especially in dismal weather. We start looking for an alternative route.

Alternative? In about eight days, we could walk out via Lukla to Jiri and then return to the GHT High Route. None of us is enthusiastic about this option and we finally consider it to be the best to try our luck. 'Good. Let's walk up the Thame valley towards Tashi Labsta. Maybe the weather will change? If not, we can still walk back and try a different route', I say. The others agree.

While my group does some necessary shopping, I leave Namche on my own to walk to Thame. I arrive a few hours later, or at least I assume I do. It starts snowing again and I look for a place to stay. One of the hotels in particular attracts me. There are flower pots on the windowsills, and the frilly curtains with floral designs that remind me of Germany; my home country. This feeling is enhanced upon entering the spacious dining hall. 'Please, take off your shoes', the strict voice of the host tells me. This goes without saying for every foreigner visiting Nepal. What surprises me most is the fact that she gives me a pair of slippers to be used inside the house.

Everything in here is spotlessly clean, tidy and well-organised which is extremely unusual for a hotel away from the main route. My host guides me upstairs to the room that evokes memories of huts in the Alps; wooden panels all along the walls, lovely curtains and brightly shining white bedcovers. This is my place for the next night, I decide.

It is already getting dark, and I wonder about the whereabouts of my team. Has something happened? Maybe they cannot find me? The latter possibility I dismiss as there are few hotels in the village. Another hour passes, and I begin to worry when suddenly the door opens and Temba enters the dining hall. 'What are you doing here? We wanted to go to Thame and meet there', he says. Totally puzzled I stare at him. 'I am here in Thame', I reply. 'No, sorry, this is Thamo', he explains with a smile. The mere thought of having to pack my rucksack again and leave this 'German' home makes disappointment build deep inside me. I suppose Temba senses my frustration because he quickly adds that the rest of the group is already on the way. They will arrive soon to enjoy an evening in this homely atmosphere.

DAY 43
THAMO – THYANGBO KHARKA
The Nangpa La Shooting

It only takes us about an hour to stroll up the valley from Thamo to Thame, and we already get there in the late morning. Our early arrival leads to the question of whether it is better to spend the night here and follow the stages according to our original plan, or whether we should try to make up for the lost time, lost through my stupidity. We opt for the latter and walk on after a cup of tea.

Thame and the Nangpa La to the north of the village have gained notoriety over time. In the past, the trail over the 5,716 metre high pass had been a traditional trade route between Tibet and Nepal, but after the Chinese occupation of Tibet, the trail began to play a prominent role as an escape route for Tibetans fleeing Chinese suppression.

In December 1999, Maria Blumencron, an Austrian producer of documentaries, was taken into custody by the Chinese police when trying to shoot a film about a group of Tibetans on their flight from persecution in China. Whereas she was released from prison two days later, her Tibetan assistant was kept in custody and tortured for two and a half years. Only a few months after her arrest, Maria Blumencron returned to finish her documentary *Escape over the Himalaya*, a heart-rending account of six Tibetan children trekking across the Nangpa La. Her work has been rewarded with fifteen international prizes.

In September 2006, a group of approximately seventy unarmed Tibetan pilgrims set out on a perilous journey. When attempting to leave Tibet via the Nangpa La, Chinese border guards fired upon them. At least sixty Western mountaineers who were at the Cho Oyu Advance Base Camp witnessed the ambush and accounts from the climbers were consistent in stating that they saw Chinese armed personnel kneel down, take aim and open fire at the Tibetans moving slowly through chest-deep snow. There were at least fourteen children in the group, and some of them were as young as six years old. Kelsang Namtso, a seventeen-year old nun, was killed and several others were injured. Thirty two Tibetans, including children, were taken into custody. Many were later released, but they had been tortured or subject to hard labour. The Chinese government initially denied the charges, but Sergiu Matei, a Romanian photographer, who was nearby as part of a climbing expedition, had filmed Kelsang's murder. It was not long before the Romanian photographer had managed to smuggle the film out of Tibet that the incident became headline news around the world. The video drew new attention to the plight of Tibetan people under Chinese rule or, more accurately, occupation. In October 2007, a year after the tragic deaths at Nangpa La, Chinese border guards shot at another defenceless group of Tibetan refugees in the same location. This time, fortunately, nobody died.

While walking on to Thyangbo Kharka through fog, low clouds and snow flurries, I cannot help thinking of Tibetans who lost their lives in futile attempts to escape the atrocities they are subject to in their home country. The situation has not improved since the genocide started at the end of the '50s, but since the economic power of China has taken off, politicians in the Western world frequently turn a blind eye to the violation of human rights.

DAY 44
THYANGBO KHARKA – CAMP
Taking a Chance

A blue sky and a brightly shining sun greet me at six o'clock in the morning and we assume we will be lucky with the weather for the whole day. In a hurry, we break camp and, half an hour later, are walking up the valley. Do we have a chance to reach Tashi Labsta tomorrow?

The first dark clouds begin to gather round about midday. At first, they obscure only the higher peaks, but soon form a solid, opaque blanket which slowly sinks down into the valley and we are enveloped in a thick whiteness – just like yesterday. The heavy snowfall of the last couple of days results in acute avalanche danger, and all day long we hear low, bass-like rumbles and witness the white death forcing its way down the steep faces of the distant mountains. We know that being cautious does not eliminate the risk completely yet, right now, we are safe in the middle of the valley.

In the afternoon, the weather becomes dismal, and we think it is best to stop and set up the tents as soon as possible. This, however, turns out to be a problem as the terrain is steep and rocky. Further up, the valley narrows, and the higher we go the closer the trail will take us to the rock walls. We know that there is one safe campsite below a huge overhang, but we do not know the exact location, or how long it will take us to walk there. By now, visibility has dropped to twenty-odd metres, and we cannot assess the safety of potential campsites through the clouds and fog obscuring the landscape. To add to our woes, snow begins to fall.

While Pimba, Sonam, Lakpa and I seek shelter from the driving snow and strong wind under a big boulder, Temba and Kinsang walk on in search for a camp. Within seconds, they have disappeared in the white-grey nothingness. Soon, the damp cold creeps into our bones and we shiver despite our thick down jackets. In the biting wind, we become ever more miserable and I try to figure out what it would be like if we had to stay here for the night; it would undoubtedly be a sleepless night of convulsive shivering. When Temba and Kinsang finally return after twenty minutes, it feels as if two hours have passed. They find a tiny flat area further up on a ridge, and through the thick flurry of snow, we follow them there. Whether we are fully protected against avalanches at the camp is difficult to decide as we cannot see anything. We can only hope.

FOUR

ROLWALING

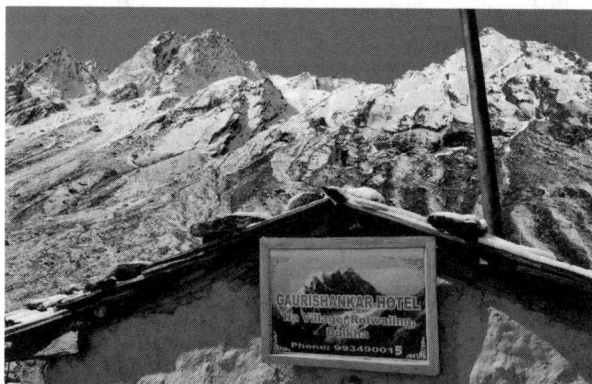

Upper part of Rolwaling Valley

DAY 45
CAMP – TASHI LABSTA – TRAKARDING GLACIER
We give it a try

After the storm rages all night along the Thame valley, I am utterly baffled to find our tents intact. What surprises me even more is that there is not one single cloud in the sky. A clear, dark-blue blanket stretches from east to west and from south to north. The weather conditions for crossing Amphu Labsta, the last technically demanding pass, could not be better. Occasionally, it pays to take a chance.

At 5:30 a.m. we eat Mount Everest-sized heaps of Dhal Bhat and, within an hour, have packed our equipment and set out on the long ascent to the pass. As usual, I try to follow a strict rhythm while plodding up the snowy slope: 100 steps and then a short rest. For some unknown reason, I do not manage today, and sometimes I have to stop after seventy steps. A general tiredness affects me, and of course, I look for explanations and excuses.

Maybe the heavy storm last night, kept me awake. Maybe the full moon that shone through the clouds was too bright and I did not sleep properly and 5:00 a.m. is not my favourite time to rise, even if I have slept well… Or perhaps I ate too much Dhal Bhat. Or perhaps too little Dhal Bhat? Could it be that there are too many things in my rucksack and it is too heavy?

I feel tired, and neither excuses nor explanations are going to change this fact, a fact that does not suit me at all, of course. Slogging up the snow turns into a fierce battle between body and mind. My experience tells me that success in many different fields depends much more on positive thinking than on anything else. Setting out on a 'mission' doubtful of a successful outcome usually means failure from the beginning. Some people deemed my plan to cross Nepal an act of folly and madness or the result of a late mid-life crisis, but I have never questioned the success of my idea; not for one single second. Now, I have to find a way to motivate myself. To avoid further disappointment and despair, I decide to adjust my strategy; fifty steps then a rest. Soon, however, I curse this idea because getting to thirty makes me realise that I will not survive the residual twenty steps. I am attentive to the symptoms of high altitude sickness, but it is neither breathlessness nor a violently beating heart that slows me down. Maybe my difficulties are just related to age? This theory does not suit me either but, in the end, I slow down even more as stopping too often would make it even more difficult to find a rhythm.

A few hours later, we face a six- to eight-metre-high wall of rock and ice that does not look particularly dangerous, but causes serious problems. The ice is melting, and as soon as we swing our ice-axes at it, it separates from the cliff and falls away. I manage to avoid a fall at the last moment when brittle water ice crumbles away and one foot loses contact. I assumed the front points of my crampons had bitten solidly – clearly they had not! It takes a minute or two for me to recover from the shock and for my heart rate to return to normal. On reaching safe ground, I need a sit-down and consider myself extremely lucky.

We meet no more technically difficult sections along the ascent route, but this does not mean that the next part is any less dangerous. Avalanches of snow, ice and stones thunder down the almost vertical walls on both sides of the basin and, witnessing the destructive forces of nature, fear creeps through my body. Yesterday, the massive rock faces had been at a 'philanthropic' distance and the rocks and snow-slides slithering down the gullies had never exposed us to any danger. Today, the walls are very close – too close. I do not feel comfortable at all.

We get to the huge overhang around lunchtime, our late arrival indicating with uncompromising clarity that we would have never made it this far yesterday. We do not have time for a long break and so, quickly, we eat a few chapattis and drink tea while snow and icicles shoot down the steep flanks beside the lunch spot or hit the ground two metres in front of us as we crouch under the overhang. It is comforting to know we are 'safe' here.

The mountain scenery is partially hidden in gloomy clouds when we arrive at the pass two hours later. In bad weather, the descent from Tashi Labsta is considered dangerous. There are no technical difficulties as such, but trail-finding on the steep and rocky slopes further down can cause serious problems. The cairns marking the trail may not be visible in dark or cloudy weather, and the likelihood of getting lost is high, entailing the risk of plunging from one of the vertical cliffs.

Fortunately, the clouds do not come closer now, and the visibility remains perfect for the rest of the day.

DAY 46
TRAKARDING GLACIER – NA
Falling rocks II

Today, my blood runs cold; more than once.

An enormous glacier and a maze of moraines form the head of Rolwaling valley. When we set out after breakfast, the trail runs safely in the middle of the glacier, which is covered by stones. The walls of the mountains surrounding the basin are steep and consist of sand and loose rock and we regularly witness rocks and stones jumping and bumping down the steep slopes – luckily far enough away from us not to cause serious concern.

The trail, however, soon leaves the safety of the glacier and forces us to walk in close proximity to the crumbling wall on our left-hand side. Now, the situation becomes serious as the mountainside is dangerously unstable, and falling stones could hit one of us any time. In order to minimize the danger, we start scrambling over the masses of scarred boulders as quickly as possible. My eyes search the terrain as I listen attentively to the rumbling sounds, trying to discover the cause of the noises and to get a clue about the size and direction of the rocks that occasionally tumble slowly to a halt close to the track. For some unknown, idiotic reason, I am convinced that knowing the course would give me a chance to escape. The terrain, however, is not the best to run on, and I would be too slow and anyway – run? Where to?

In the end, I succeed in simply ignoring the potential danger by repeating to myself, 'Not me! Not today!' This may sound silly, but it helps me to overcome the paralyzing fear that rules my thoughts. To my immense relief the walls of the mountains 'retreat' as the valley widens three hours later. My tense muscles relax, and it is only then that I am able to take pleasure in the magnificence of the landscape.

After a long and tiring walk, we arrive in the village of Na that is entirely in Norwegian hands tonight. All foreigners (four) are from Norway: a young woman, doing some ethnological research; a doctor and his assistant, testing the effectiveness of nitrate on the performance of the human body at high altitude and me.

DAY 47
NA – DOKHANG
Drinking while 'on duty'

New snow fell last night, and it is cold when we leave Na. I do not envy Karoline, the student, who will stay in the village for a few weeks more. Or maybe I should say 'has' to stay?

The traditional houses of the Sherpas are built of roughly cut stones, and I deem the buildings neither to be homely nor warm places in winter. Heating systems as such do not exist, and this explains why family life is confined to the kitchen, which usually occupies the entire second floor. The kitchen is not only a place to cook and eat, but also an area where people sleep, work and pray; where children play and do their homework whilst sitting on the floor beside the hearth. Since the guest room is as cold as a fridge, Karoline's bed is also in the kitchen.

Today's journey is a lovely stroll down the valley and with every hour that passes, the temperature rises. The down jacket and the warm pair of trousers are the first clothes to disappear into the rucksack, followed by the woollen jacket some time later. Around noon, we get to the village of Beding where a few basic hotels cater for the needs of occasional tourists. Kinsang disappears in one of them. I cannot help but suspect the reason…

On the journey across Solu-Khumbu, Kinsang stayed behind regularly, and I often wondered why. It did not take long to find out that he spent his time (and his money) on drinking. I do not object to alcohol in general, and now and then, I enjoy drinking Tongba or Chang in the evening, but I object to drinking during the day, drinking while 'on duty' as I call it. Kinsang and I had discussed this issue before.

I decide to walk over to the hotel which Kinsang has disappeared into, and my suspicions are confirmed, unfortunately. The talk, meant to clear up the situation, soon turns into an agitated argument. I suppose it is worse for him that the criticism comes from me, a woman. Maybe, it would have been tactically more prudent to send Temba to the 'front'. Now, it is too late.

The lunch break ends with Kinsang totally drunk. He staggers and stumbles along the trail and Temba has to lead him by the hand all the way down to Dokhang. Thus, we are sure that Kinsang does not plunge into the river that runs near the trail. I have already moved into the hotel, made myself comfortable and ordered some tea, when the two finally arrive; Kinsang singing and mumbling incomprehensible gibberish.

DAY 48
DOKHANG – CHETCHET
I can hardly believe my eyes and ears

When I wake up, the warmth of the sun greets me. It was only twenty-four hours ago that we experienced new snow in Na and, in my opinion, the contrast between the cold we left yesterday and the green surroundings at Dokhang could not be more marked. We are at an altitude of approximately 2,800 metres and thus 1,400 metres lower than Na. According to a rule of thumb, 100 metres difference in elevation mean 0.8 degrees difference in temperatures; we can enjoy a plus of eleven degrees!

Today, the trail runs downhill, taking us closer to the summer with every step we make. Once again, I am surprised by this sudden change. The enormous differences in altitude in the Himalaya offer the unique opportunity of experiencing all four seasons within two or three days. Yesterday in the morning, we started in the middle of winter, today we woke up in spring and we will probably arrive in summer this evening.

We agree on meeting in Simigaon, and I set out on my own. It is elating, being surrounded by green trees and bushes, hearing the singing of birds, finding the first flowers of the year and feeling the warm wind touch the bare skin of my forearms. Since we left Hatiya more than three weeks ago, we have travelled in winter conditions most of the time, and I have sporadically thought longingly of spring time in Norway, which is the best time of the year. With the days getting longer, nature seems to explode overnight, finally escaping the strangling grip of ice and snow. Today's walk is a compensation for missing out on spring back home.

When Simigaon comes into sight, I can hardly believe my eyes and ears. Is there a car blowing its horn somewhere in the distance? Is there a sandy trail beside the river or a road? Soon, I know that neither my hearing nor my vision is impaired. There is a road with cars and lorries driving to and fro, horns blasting. After about fifty days far away from civilization, these noisy metal boxes on four wheels are a shock and I am able to visualize the astonishment, or fear, the Nepalese people felt when the first car drove through Kathmandu in 1940. Since roads were lacking then, porters had to bear the vehicle over the mountains all the way from India. The car, a 1938 model Mercedes-Benz, was Hitler's present for Juddha Shumsher Rana, the prime minister of Nepal at that time. Of course, people speculated on Hitler's reason for sending such a precious gift and suspected Hitler of trying to buy the Gurkha soldiers' support. He wanted them to fight for Germany in WWII instead of joining the army of the British Empire.

In Simigaon, I have a long conversation with the headmaster of the local school who also runs one of the hotels in the village. He is interested in the work of Autism Care Nepal, and I try to give as much information as possible, but the topic is too complex to be explained in one or two hours. When he tells me that he is in charge of the training of young teachers, I hand over a pile of brochures about autism for his students. Further talk shows that he had been in charge of several outstanding projects in Simigaon: a bridge, which will save the locals a long detour; a health post; a gompa and a community lodge. The latter has not been finished yet, but in the near future it will, hopefully, generate income for the community and create working places for young people. His latest project is the construction of two bathrooms with showers besides the school building. None of the local families' houses are equipped with sanitary or even shower facilities and he is tired of dirty children coming to school. 'This has to stop', he adds.

Afternoon has almost passed when Temba turns up in Simigaon. He informs me that the rest of the group had taken a different route, leading directly down to Chetchet. I would prefer to stay here since the view from the ridge is astonishing, but maybe there are better hotels and even electricity down in the valley…

We drink a last cup of tea with the headmaster and his wife before descending. In the beginning, the zigzagging trail passes a few scattered houses and runs beside well-maintained and terraced fields until it eventually winds through a forest. It is only then we notice the darkness that has arrived. The trail resembles a faint irregular line in front of us but, cautiously feeling our way down, we arrive safely at the brand-new concrete staircase leading to Chetchet.

One member of the group has not arrived in Chetchet: Kinsang. He had left his crampons somewhere along the trail and had turned back. He will come later.

DAY 49
CHETCHET
A rest day that was not planned

If you ever need to plan a rest day on the way out from the upper part of Rolwaling valley, do not choose Chetchet! We had to, and so I know what I'm talking about.

Kinsang did not arrive at the hotel last night and today the rumour goes that he began to suffer from some serious stomach problems and had to stay behind at one of the lodges in Simigaon. We are concerned about Kinsang, but since there is no mobile coverage in the narrow valley Temba and Pimba decide to walk back to Simigaon.

I try to make the best of the situation, washing some of my clothes, arranging pictures on my computer and updating my diary while enjoying the warming rays from the sun. This doesn't last long! It soon gets hot and I desperately try to retreat to the shade – only to find that there are neither trees nor bushes and that the hotel is not a pleasant

Hotel owner's son

place to escape to as it's too dirty. I fantasize about ice cream and cold drinks (although, without electricity, this must remain a dream...)

Right in front of my tent, trucks and construction vehicles drive to and fro, leaving behind offensively smelling clouds of dust. The Chinese are not only building a road in this part of the Rolwaling valley, but also a giant hydro-electric power station, which will provide electricity to the whole region within one or two years. I'll probably be able to get a cold drink next time I'm here – although I doubt that the dreariness of the place will disappear. Everywhere are piles of rubbish, empty plastic bottles, old shoes and torn clothes, soup packages and innumerable beer and whisky bottles. In most Nepalese villages, there are neither waste disposals nor recycling systems. What do you do with things one no longer needs or wants?

In the early afternoon, we are still without any message from Kinsang, or Temba and Pimba, and since we are seriously concerned, Sonam decides to walk back to Simigaon to look for them. I would not want to step into his shoes, even if offered 10,000 NRP, as the steep slope is exposed to the sun. I assume that following the zigzagging trail is like walking inside an incinerator.

They return just before dusk: Temba, Sonam, Pimba and Kinsang, who staggers along the road and needs help to walk up the few steps to the fenced garden in front of the hotel. His eyes are red, and he cannot remember what happened during the last twenty four hours. Later I learn that locals had found him somewhere in the forest, completely disorientated. This does not look like a stomach problem... After a quick 'emergency summit' with the rest of my team, I declare Chetchet to be Kinsang's final destination. Tomorrow, we will put him on a bus to Kathmandu and, to make sure that he arrives there safely, Lakpa will join him. I pay Kinsang the salary we had agreed to. End of discussion.

DAY 50
CHETCHET – BULUNG
Chinese Presence

The village of Gonggar, from where buses leave for Kathmandu, is less than one hour away from Chetchet. Since nobody in Chetchet knows the time of departure, we get up early and start walking along the dusty road at six o'clock. Kinsang still suffers, and so we decide to go ahead with the original plan, Lakpa will accompany him. Of course, we are sad

Chinese presence

to lose our cook, but his journey back to the capital provides a unique opportunity to get some culinary delicacies, and so we write a long shopping list for Lakpa to take with him. We agree to meet again in about five or six days at The Last Resort, a hotel situated on the Araniko Highway, also known as the Kodari Highway.

A walk through Gonggar evokes impressions of roaming through a village in China. It is filled with hundreds of Chinese labourers – all wearing blue suits and red helmets – who are boring a tunnel for a massive hydro-electric power plant. There are Chinese restaurants with red lanterns swinging gently above the entrances and shops that offer almost exclusively Chinese goods; warning signs and information plates in Chinese; cars and lorries with Chinese licence plates; enormous heaps of rubbish mainly consisting of broken Chinese products and thousands and thousands of empty bottles with Chinese labels. What causes the Chinese to be so actively involved in co-operation and charity projects like this plant? Up to now, nobody has provided a satisfactory answer to my question.

The roar of heavy machines reverberates through the valley and is amplified by the steep mountains, increasing the volume to an unbearable level. I long to be back in the undisturbed, comforting peace and tranquillity of the high mountains, with only the sound of the wind and the flapping of prayer flags.

We continue along a track that takes us uphill to a ridge and, finally, away from the 'valley of horror' with its deafening sounds. The contemplative walk back to the unspoilt countryside does wonders to my stressed mind and I am able to fully appreciate the approaching summer. The undulating hills resemble a painting, showing every shade of green an artist could imagine. It is scenes where a camera is useless, unable to capture the intensity and power of the colours that nature displays. The wind carries the scent of thousands of flowers and the gentle sound of bees buzzing from one blossom to the next. In my mind's eye, I conjure up summer holidays from my childhood when I used to stroll across the fields behind our house with my friends from school. We had neither plans nor appointments, we would decide our destination on a whim. There were exciting bike races along rough sand roads and excursions into the nearby forest, (where, needless to say, we were not allowed to go…). We hid in the waist-high grass and pretended to be Red Indians hunting buffaloes. Countless memories come and go while walking, like waves rolling gently in and out a sandy beach.

We have arrived in a Nepal for Nepalese people; a Nepal where strangers are met with both open curiosity and wary scepticism. On arriving in Bulung, we start searching for a potential campsite. Within half a minute, we are surrounded by a big group of people who follow us all the way through the village to a dilapidated hotel where we decide to set up the tents. The villagers – old and young – are everywhere, observing every step and move we take or make. As soon as the kitchen tent is ready for use, they start crowding in to examine our belongings. I do not feel comfortable in this situation and disappear in my little red home, closing the zipper behind me.

Locals pay us a visit

DAY 51
BULUNG – LOTING
EcoHimal project

When I wake up at a quarter to seven, the locals are already back; squatting in front of the hotel and waiting for us (or just for me?) to crawl out of the tent. All I want is to eat my breakfast undisturbed and without small talk. Mornings are not a time of the day when I am particularly social.

The sun has arrived along with the locals and the heat soon forces me to open my tent. Obviously, that is exactly what the locals want, and I suspect that there is a secret agreement or contract between the inhabitants of Bulung and the sun. As soon as the zipper slides down, they gather round and stare at me. Their curiosity collides with my need for solitude and I feel anger building up inside me. As a result, I get angry about my anger because I do not accept that strangers like me are a substitute for TV entertainment in a remote village like Bulung. But how would I react if I were in their position?

One hour later, we are back on the trail, crossing green, rolling hills again. Only five days ago, we strolled through a country that was still in the grip of winter, and here the wheat is beginning to ripen, and even be harvested in the sunnier places. The contrast is enhanced by seas of swaying flowers lining the path and hundreds of butterflies in all colours and sizes.

The Gauri Sankar/Rolwaling differs from trekking areas like the Solu-Khumbu, Annapurna or Langtang, where tourism development has a long history. For the local communities in this area, tourism is a supplement to their subsistence farming, not a primary source of income.

South of the holy mountain of Gauri Sankar, the Austrian organization EcoHimal operated an ecotourism project from 1996 to 2008 aiming to create new jobs and thus reduce migration to cities and to foreign countries. The project was based around the notion of making tourism planning compatible with local culture and environment. A controlled development of tourism is a valuable contribution to sustainable regional development and results in improved living conditions for the local population.

The first step was to create the necessary tourism infrastructure. Together with the local people, the Austrians built drinking water systems, lodges and campgrounds and improved already existing trails and bridges. Since almost all the villagers lacked tourism-related experience and knowledge, EcoHimal held workshops and training courses to educate and prepare them. At the end of 2008, the project was completed, and the local communities took over the management of lodges and campgrounds. The idea as such was brilliant, but reality was disillusioning. What happens when a group of people shares responsibility and the generated income? *Who* is responsible for *what*?

Of course, I have not seen all EcoHimal Community Lodges, but those I passed on my trip were run-down and the toilets and showers rarely worked properly, if at all. For a couple of years, the region has been under the spell of the power plant, and ecotourism has been put on the back burner. Local people prefer focusing on building roads and hydro-electric power stations. In addition, local politicians have undermined and boycotted the project in favour of their own plans. Marketing worked only as long as the Austrians were present. Now they are gone, trekking agencies focus on marketing tent trips because their costs are easier to calculate and the rare individual trekkers visiting this area do not provide a high enough income for the local communities to maintain lodges and other tourism-related facilities. With the Austrian government no longer allocating financial resources to the project, it is questionable whether the lodges can now ever be 'saved'.

A sobering reality which, once again, shows that aid projects without controlling and supervising donor presence often fail; even though they are well meant. We, the foreigners, often underestimate local structures and the fact that other countries may have different priorities since their needs are different from ours. It seems that, generally, the individual comes first – both among locals and among travellers, who often come for the high peaks; not for rural culture.

The quality of the lodge in Loting makes me opt for my tent tonight.

DAY 52
LOTING – BIGU GOMPA
Heatstroke

Because of the ever-worsening heat in the 'lowlands' (areas below 3,000 metres), we leave the EcoHimal lodge early and continue the journey towards Bigu Gompa. The early start, however, does not help, and the heat soon begins to wear me out. The sun glares down, the suffocating heat seems to sap my energy and a collapse feels close. I get a splitting headache, my stomach revolts, my heart rate is somewhere close to 180, and the dizziness tells me that my blood pressure has dropped significantly. These premonitory signs of heat stroke are familiar to me and it wouldn't be my first. Since we arrived at lower altitudes a few days ago, I have taken any possible precautions to avoid such problems; wearing a hat all the time, cooling my head and sprinkling my clothes with water at least every other hour. The sudden rise in temperatures, however, is simply too much for me.

When I finally crawl the last few metres to a village situated on top of a ridge, I see my group disappearing behind a building, but am too fatigued to shout for help. There is hardly any energy left in my system when I totter to one of the houses where the roof overhang promises protection from the sun. This is the end; my end.

Time passes, and over-curious people gather round to gawk at me. Pupils from the local school, which is only thirty metres away, come over and start flinging small pebbles at me, possibly trying to discover whether or not I am dead after I fail to respond to their comments and questions. All I manage is wondering about how the locals survive in these tropical temperatures. They walk around with thick jackets on, woollen hats covering their heads and scarves wrapped around the neck.

I finish the last drops of water from my bottle, and suddenly, a wave of hopelessness and discouragement grips me. I fear I will be unable to endure the parching thirst and scorching heat long enough for my team to notice my absence. My perception of time is clouded, and it seems to take an eternity for Temba and Sonam to return to the village and to my rescue. When Sonam insists on carrying my rucksack, I am too weak to resist, mentally and physically. Temba tries to find drinking water or a shop selling soft drinks, but there are no shops in the village and for safety reasons I decide against the water from the public water tap. The lunch place is less than twenty minutes away from the village, but for me, the walk turns into a nightmare, making me want to stop, to sit and to die. Fortunately, my death is prevented by Pimba who had already prepared some tea for me. I drink several cups in a row before falling fast asleep on a mattress. Today, it happens for the first time that I leave the Dhal Bhat untouched, and so everybody understands that something is wrong with me; seriously wrong.

Waking in the afternoon, I still feel miserable, but since higher altitudes equates to less heat, I am determined to continue to the monastery at Bigu. Soon, however, I regret my plan, but as the only alternative would be to walk back down (an idea I like even less), I begin trudging up the hill.

On reaching Bigu Gompa, I disappear in my tent without dinner, freezing and shivering.

DAY 53
BIGU GOMPA
Another rest day that was not planned

After a twelve-hour sleep I feel slightly better and even tinker with the idea of continuing the journey, but walking ten metres to the bathroom tells me that these ten metres are my maximum. We decide to have a rest day at Bigu Gompa; unplanned.

To be quite honest, I hate Coca-Cola and doubt I have drunk more than twelve bottles in the whole of my life. In my opinion, the drink is pure poison (like medicine). But, when feeling sick, taking medicine becomes a necessity and so I declare the day a 'Coke medicine' day. Luckily, the shop at the Bigu Gompa Community Lodge has Coke on stock – enough to cure hundreds of sick people. I lie down on the balcony of the building and begin to sip from a bottle, then another, and another… Having emptied three bottles, I feel strong enough to walk to the monastery, 200 metres away from our camp. The success of my self-medication fills me with pride and joy.

Bigu Gompa is a nunnery currently housing more than seventy nuns. A school for girls is part of the convent, and as it is well-known all over Nepal, parents from across the country send their daughters to Bigu Gompa to become nuns. The majority of the girls come from poor families and sending them off means fewer financial problems at home. I ask one of the older nuns for permission to take pictures, but she objects to taking any inside the monastic buildings. 'You can take pictures of the young novices if you want to', she tells me. This generous offer is a worthy compensation but, unfortunately, many girls are shy and hide their faces in the folds of wide maroon robes, occasionally giggling and laughing.

The novices get a basic general education, but the standard schooling, including subjects such as mathematics and English, stops after grade five. The English classes, in particular, depend on volunteers from abroad who have enough time at their disposal to stay and hold lessons because the government does not send teachers and the monastery lacks highly-educated nuns for this job. After the basic education, the girls will continue reading and studying only religious texts. Usually, the situation is different in monastic schools for boys where teaching general subjects will continue after grade five, providing education equivalent to standard schools.

Why is there a difference? This question can be answered easily. The eight Garudharmas (strict rules) place nuns below monks and consequently nuns are generally considered inferior throughout the Himalaya. Nuns are

Girls at the nunnery

systematically kept away from sophisticated work and are more likely to be placed in jobs like cleaning and cooking. Men dominate institutionalized religion – even at Bigu Gompa, two lamas are in charge of the monastery.

Regulations like these show clearly that Buddhism is not free from gender-based discriminating, or at least that it is common practice to turn a blind eye to it. The words uttered by Sayadaw U Asabhacara from the International Buddhist Meditation Centre on the 26th of November 1991 show this clearly:

'Women by nature are not powerful, both in body and mind... When women get power they become proud... It's a natural happening for men to have control over women... Buddha's preaching is very fair'.

Luckily, a new way of thinking is beginning to enter Buddhism and 'revolutionary' ideas are permeating even the dust-covered monastic life of nuns. Or, at least, they are in Kathmandu valley where the 800-year old Drukpa Buddhist sect broke with traditions, nuns are usually taught by learned monks, but His Holiness the Dalai Lama sent four experienced Vietnamese Drukpa nuns to serve as teachers. In contrast to most Buddhist sects, the Drukpa nuns learn to lead prayers and receive a basic training in the business skills required to run the guest house and coffee shop at the monastery. They can also take driving lessons... after which off they go by jeep to Kathmandu to do the shopping for the convent.

The nunnery even offers martial arts training – after the introduction of kung fu two years ago, the popularity of the nunnery soared. At the moment, a Vietnamese master trains about 300 young Buddhist nuns, who practise kung fu fighting for up to two hours a day. One of the aims is to make the girls and women more self-reliant. Eighteen-year old Jigme Konchok Lhamo, who came from India to the monastery, explains that kung fu has made the nuns more confident and helped to alter the power balance between men and women in Buddhism. 'His Holiness wants the nuns to be like the men, with the same rights in the world', she says and adds: 'That is why we get the chance to do everything, not just kung fu. We also have the chance here to learn many things, like tennis and skating. And we can learn English and Tibetan and musical instruments'. Bigu Gompa is still some distance from Western ways of thinking, but its style of changing monastic life is playing a vital role in women's and girls' emancipation.

Culture and tradition have ruled – and still do so – the life of women in Nepal with hard work in the parents' home, an early marriage, and then hard work in the husband's home before giving birth to children (as many as possible – which results in even more hard work). Self-determination over one's life is a privilege only some women from a few ethnic groups have. Monastic life offers an opportunity to escape the toil and the hardship culture and tradition demands. Ironically, being subject to the strict rules of a nunnery equates to almost unabridged freedom for girls and women.

By the time the sun goes down, my state of health has improved and my well-known appetite is back. Temba is happy when I come to the kitchen tent and ask for Dhal Bhat. This is a good sign, everybody confirms.

DAY 54
BIGU GOMPA – TINSANG PASS
Gold fever

Although I still feel weak, we decide to push on and to leave Bigu Gompa. Luckily, the path is not steep, but I still tire quickly, and would prefer to sit under a tree rather than trudge up the sandy trail to Tinsang Pass. All day long, I have to push and motivate myself by repeating my personal magic mantra: 'Do not give up that easily'. Nevertheless, it comes as a relief when we arrive at our destination in the early afternoon.

The hamlet, situated in the middle of nowhere, consists of three small, basic hotels beside the dusty road and the prevailing atmosphere of the place resembles a scene from an old Wild West movie: dust dances in the air, drunken men stumble around waving empty bottles and screams and shouts accompany their card games. To make sure that we remain undisturbed, we pitch tents a few hundred metres away from the buildings, just in case. One never knows…

The GHT guidebook describes the following day as having complicated route finding and recommends employing a local guide. According to my experience, locals are generally aware of the fact that a tourist depends on their knowledge and, therefore, they often claim some 'extra' money. In my opinion, this is fair enough, and since this 'extra' will help us to save time I am willing to offer a double wage. This always works. Always? No. Today it does not. When Temba returns from the hotels where he had tried to find a local guide, he tells us that 7,000 NRP had been the lowest price one of the men had asked for, approximately 70 Euro! I am speechless, but an explanation for this astronomic wage claim is close at hand.

Here, the modern Nepalese 'Gold Fever', the 'Yartsa Gunbu Fever', dominates the people's lives. In rural areas of Nepal, Yartsa Gunbu has become the main source of income and rumours about earning 'dough overnight' are everywhere.

Yartsa Gunbu is a fungus which grows only at altitudes between 3,000 and 5,500 metres. The fertile, sub-Alpine slopes of the Tibetan Plateau and the Himalaya are thus ideal for it to thrive. The Tibetan name Yartsa Gunbu can be translated as 'summer grass, winter worm' and, scientifically, the fungus is neither a plant nor an animal. It grows from the larva of a ghost moth which dwells underground where it is often infected by spores from a parasitic fungus called the *Ophiocordyceps sinensis*, which devours the caterpillar's body, leaving only the skeleton intact. In spring, a new fungus grows out from the head of the caterpillar.

According to specialists, the price for 500 grams of top quality Yartsa Gunbu increased by 900% between 1997 and 2008, with people paying 13,000 USD in Lhasa, Tibet and up to 26,000 USD in Shanghai. But why is Yartsa Gunbu so valuable? Yartsa Gunbu was first mentioned in a fifteenth century text and was described scientifically in 1843 by Miles Berkeley, a founder of the science of plant pathology. For centuries, practitioners of Tibetan medicine, Chinese medicine and traditional folk medicines alike, thought Yartsa Gunbu to possess magic-like medicinal and libidinous powers. It is said to cure a variety of ailments: hepatitis, bronchitis, back pain, impaired vision, asthma, HIV/AIDS, cancer – and more! As Yartsa Gunbu is apparently both animal and vegetable (actually it is not vegetable, but fungi), people believe it to balance the effects of Yin and Yang, thus improving the state of health in general. Even yaks are reported to benefit from the miraculous power; they grow in strength tenfold.

The 'Yartsa Gunbu Fever' breaks out in spring and comes to an end with the onset of monsoon. During the harvest season, ten thousands of people crawl quietly across the sub-Alpine pastures, searching for the fungus. Finding the elusive 'Gold of Nepal', however, is difficult and requires experience and excellent vision (the visible part of the fungi is the size of a match and easily mistaken for a blade of grass). Days may pass until a searcher finds one single Yartsa Gunbu, and he will demonstrate his joy over the find by crying out in pleasure and excitement, causing dozens of other people to run over to him to witness his luck.

A fortunate harvester may earn as much as 150 Euro in one single day. Anything rare has a value and attracts money. A lot of money – especially when all attempts to farm the fungus have failed and this fact explains the high price.

Similar to the gold diggers' mining rights, harvesting areas are strictly regulated but, due to the incredible value of Yartsa Gunbu, local communities report increasing inter-village quarrels over harvesting rights. In June 2009, inhabitants of a remote village in the northern district of Manang killed seven farmers who had collected Yartsa Gunbu 'illegally' in an area owned by the local community. Two years later, in November 2011, nineteen villagers from Nar/Pho were convicted by a Nepalese court over the cruel slaughtering of a group of men. They had been fighting over the rights to collect the precious aphrodisiac fungus.

Police in Dolpa expect 40,000 people to migrate to the district this year. The unprecedented flood of harvesters and the environmental impact are not the only concern and there are fears of quarrels between the Dolpopa and strangers who have to pay a 'harvesting fee' to the local communities – particularly when many people try to circumvent this regulation and collect the fungus illegally. The police stations there are fully staffed.

In the hope of finding a local guide, Temba walks over to the hotels once again, in vain. At least, one of the 'gold miners' gives him a quick description of the route, free of charge.

DAY 55
TINSANG PASS – LAST RESORT
Enchanted Forests

Walking towards forested areas in Nepal, locals often warned me of 'bad' people and usually had two or three stories at hand: a business man had been robbed, a trader killed, a woman molested and so on. Yet I had never experienced any situation I perceived as threatening and dangerous. Are these stories folklore only? Even nowadays, the people in Nepal live close to nature, and their scary stories are, probably, the result of reality, fear, superstition, lore and respect for the unpredictable forces of nature.

All around the world, wherever forests abound, legends and fairy tales of enchanted forests are common. Forested areas are either places of danger, refuge, or adventure. Hänsel and Gretel met a cannibalistic witch in the forest, Brother and Sister found refuge in the forest after their stepmother had turned the brother into a deer, and the story 'The Three Little Men in the Wood' describes adventures in a forest.

In a few cases it is not the forest that is enchanted, but the creatures, plants, rocks and creeks in it. These magical forests are full of talking trees, vicious branches (reaching out for you or nudging you off your horse), thorny bushes (allowing you into a magic cave, but not out again), plants (turning into animals at night) and creeks (turning unwary travellers into frogs). At night time, elves and fairies dance joyfully to eerie music in the clearings and, last but not least, a forest and its dense undergrowth are the perfect habitat for sorcerers, giants and dwarves – or just a good hiding place for robbers.

Finding the way up to the pass is easy, but we face serious problems when walking down the other side. There are trails in abundance and the directions Temba received are not helpful at all. A couple of times the chosen path ends in the middle of nowhere and we are forced to turn around, climbing back up again to look for a different trail and a different line. We stay within earshot of one another – losing touch would be disastrous. A wave of fear runs through me when I think about getting lost in this maze of densely-forested valleys. During the day I would be able to cope with being lost, but at night... Strangely enough, this fear seems to sharpen my senses since it is only now that I notice hundreds of different sounds: cracking twigs, rustling leaves, screaming apes, hammering woodpeckers, humming bees, chirping crickets, babbling creeks...

Hours later, as we leave the forest behind us, we are surprised to see some shepherds' huts right in front of us which seem to be inhabited. Smoke leaks through roofs made of woven mats and I hear a dog barking fiercely. At one of the huts, a young woman offers us tea and we learn that she and the other shepherds have arrived here only the day before. What a coincidence! How lucky we are! She describes the way down to the Araniko Highway in great detail, and we are sure to get there without encountering any further problems.

Elated, we hurry down the grassy slopes and find a suitable place to lunch in the sun. Today's highlight is a rare but delicious dessert: wild strawberries. The entire slope is full of sweet, red berries and we are reluctant to leave this culinary paradise, but time is running short, and it is still a long way to The Last Resort.

Resort is an appropriate name for the hotel we reach just before sunset. Passing through a guarded gate and entering a vast park, one arrives in a dreamland, hidden from the rest of the world. Tent cottages are snuggled into the forest. Servants and cleaning crews hurry along the spotlessly clean gravelled trails winding through the resort. There is not a piece of paper nor rubbish to be found anywhere; a litter-free zone.

With perfect timing, I have just finished taking a refreshing shower when dinner is ready, served in a restaurant with a twelve-metre long, open-air buffet which forms the core of the place. After almost two months roaming through the Himalayan mountain wilderness, I have no problems eating my way through the entire menu, or through the twelve-metre long array of delicious food.

FIVE

HELAMBU/LANGTANG

Village in Helambu

DAY 56
LAST RESORT – BARABISE
Modern tourism in Nepal

I wake to the sounds of the jungle, seemingly thousands of birds are singing, whistling, peeping, chirping and croaking. To make the jungle atmosphere perfect, I expect giant spiders to drop from the ceiling and snakes to curl round the iron bedpost. Luckily, this is not the case, although I feel sure they exist.

The breakfast is superb, and the staff members watch me with a mixture of scepticism and disbelief when I collect gargantuan servings of fruit salad with yoghurt and honey from the buffet, consume several bowls filled with muesli and finally load a couple of fried eggs and multiple of bread on to my plate. Who knows when I will get the next opportunity to enjoy such a treat?

The Last Resort is popular amongst 'modern' tourists who come for quick fix adventures like bungee jumping and rafting, tightrope team building, jungle adventures and canyoning. I feel like a leftover from another time, a time when people still walked up and down the mountains. I suppose that the majority of the guests have never walked further than to their car or the nearest bus stop.

Apparently, The Last Resort also attracts young Nepalese people who can afford to take a short break from work, jumping onto their motorbikes or taking a scheduled bus from Kathmandu to enjoy the luxury of the hotel. They come from well-off families, are highly educated and have reasonably good incomes and iPhones in their pockets. Some older people stay as well. They come for relaxation and leisure, spa treatments and short day trips. All in all, it is a fascinating mix to observe.

Yesterday, Lakpa had arrived with the bag full of delicacies we had ordered from Kathmandu – including instant cappuccino. The team is complete again, and we are ready to set out for the fourth part of our Nepal traverse: Helambu to Langtang. Still, the exact route has to be determined, but since I want to see the holy lakes of Gosainkund we will have to follow the GHT Cultural Route for a few days before heading north again; back to the GHT High Route. We leave The Last Resort and head south for Barabise.

DAY 57
BARABISE – DHUSKOT
Two years ago

Yesterday in Barabise, I purchased an umbrella and today, at seven o'clock in the morning, I am already grateful for the little shade it provides. The path is exposed to the sun and the umbrella protects me against the sun rays but not, unfortunately, from the heat and I already doubt whether it was such a good idea to put the holy lakes of Gosainkund on top of my 'to-do' list. There would have been a different, cooler route high up in the mountains, but it is too late now to change our itinerary and so we struggle up the hill in the pre-monsoon heat.

When we reach the small village of Dhuskot, it does not take long before we are surrounded by the curiously-staring inhabitants of the settlement, staring first of all, at me, I suppose. Despite their obvious poverty, the people are proud and full of dignity, and their genuine interest is unobtrusive and accompanied by smiles and laughter; I do not feel abashed at their attention. One local woman approaches me and explains – with a lot of gestures – that she wants me to take a picture of her. Usually, the local population is not delighted at all about tourists' enthusiasm for taking pictures of them, and thus her desire takes me by surprise and I am totally baffled. The woman straightens her clothes, brushes back some strands of hair and proudly poses in front of my camera lens, smiling

A brave village woman

broadly – while the rest of the female population stares at her, expressing a mixture of admiration for her courage and scepticism about her audacity.

When Pimba sets up my tent, the villagers are apparently surprised to see that only two poles are needed, and by the look of it this fact becomes the main topic of conversation for the following ten minutes. As soon as everything is in place, they come closer to examine the fabrics and the zippers minutely, check the thin strings for durability and pass one of the lightweight tent poles from hand to hand. The highlights of the 'show', however, are my self-inflating sleeping mattress and my down sleeping bag which is totally useless in the heat, by the way.

In the general hustle and bustle I notice a teenage boy watching us from the window. Whenever I cast a glance in his direction, his head disappears, but as soon as I turn away, his head pops up again. This game goes on for a while. It is obvious that he is very interested in everything that is going on, and my mattress seems to be his favourite. For some reason, however, he does not dare leave his house. Okay, I will take my mattress and pay him a visit. On entering the dark kitchen, my eyes need time to adjust to the darkness, but then I spot the boy in a corner beside the window. His facial expression reveals surprise, curiosity, wariness… I inflate my mattress, deflate it again and hand it over to him for closer inspection. He looks at me, smiles shyly and hands it back after a minute or two. When I leave the room I can watch him dragging himself over the floor back to the window. Like his sister who sits at the far end of the room, he suffers from (what I assume to be) *Osteogenesis imperfect* (brittle bone disease). Neither the boy nor the girl has ever left the village, and I notice that I avoid thinking about their future. They do not have sisters who may take care of them after their parents' death.

Sonam and Temba chat away with the headman of the village and later tell me that Dhuskot saw its last visitors about two years ago. The locals still remember the elderly couple from Australia since they had been the first tourists ever to arrive here. I wonder whether it had been Robin Boustead and his wife on their reconnaissance tour of the GHT Cultural Route.

DAY 58
DHOSKOT – BATASE
Welcomed intruders

Nepal has been without a constitution for about four years and the people are sick and tired of the delaying tactics employed by the government; a government that seems unable to get their act together. Long-lasting strikes have paralysed the country and, since public transport is affected most, the overall supply situation in many parts of Nepal is critical. As a result, our stay is affected because, since leaving The Last Resort, we have been without kerosene. In the search for fuel, Temba and Sonam had roamed in vain through all the shops along the Araniko Highway. Nobody knew whether it would take one day or even one week for new supplies to arrive from the capital. Even in Barabise, a middle-sized town with all sorts of facilities and shops, we could not fill the empty canister. What should we do in this situation? Cut branches? Fell trees? We try to avoid this at all costs.

Imagine the following situation in your home town:

All morning you have toiled in your garden and now your family is waiting indoors for you to prepare lunch. All of a sudden, you see a group of five walking along the road. They stop outside your property and only seconds later, enter your garden. Judging from their clothes, they are not from your area, and one of them is obviously from a far-away country with a different culture. What are these people up to? They lay down their heavy loads on your terrace and make themselves comfortable there! And then one of them walks right into your kitchen, gazing curiously around, and asks whether it is possible to use your oven because he and his friends ran out of fuel and would like to make some food in your kitchen. That's the last straw!

For two days, that group of five has been us – and not once did the locals turn us down when asking for help. Of course, we offer payment and usually buy a few things from our hosts; a few eggs, some rice, fresh vegetables or berries – which are a luxury found in abundance at this time of the year. These visits are memorable experiences for me as they provide insight into the normal life of the hard-working people who struggle to make a living in this beautiful but less affluent part of the country, far away from crowded trekking areas and famous tourist sites. Like the majority of Nepalese people, the inhabitants of this area are peasants, living on what they grow in their fields and the small income generated by selling animals or animal products. Though the hills of Nepal are fertile, the fields barely provide enough food for a family to make ends meet for twelve months and the little money they, (if lucky), earn in addition does not much help to make life easier. This fact explains the large number of people migrating to India, South Korea, Arab countries or even Europe – sometimes just for a few months, sometimes for one year or for good.

As we leave the settlement, we see people working in fields which, according to our host, do not belong to the labourers. In fact, many Nepalese people only own a little plot of land; some are even without any property. The fields they work on, and have worked on for generations, belong to absentee, wealthy landlords: a complex cultural and social system. According to the local NGO, Federation of Nepal:

'Most of the country's land is controlled by the Nepalese elite. As absentee landlords, most do not even live in the villages where they legally own land, yet they reap most of the income. With no legal claim to the land, or registration that even recognises them as tenant farmers, instead of receiving a fifty per cent share of the harvest, labourers get at most one-third and, more often, one-tenth of earnings'.

It is discomforting for me to observe hereditary vassals toiling for feudal masters they, probably, have never met.

DAY 59
BATASE – OKHRENI
Internet in the forest

Since parts of the trail will run uphill today, we opt again for an early start to avoid the heat that sets in early, at about eight o'clock. Depending on the terrain (up or down) and on the vegetation (forest or open plains), we walk until eleven or even twelve o'clock and then take a break until the late afternoon waiting (and hoping) for the temperatures to drop.

The pre-monsoon drought has hardened the soil, creating cracks that run in all directions and make the land look like an abstract drawing on yellowed paper. Layers of hot air above the sand and stones create inferior mirages and the blurred, shimmering effect obscures details of the landscape. Wind is said to lessen the occurrence of heat haze, but no cooling breeze blows down the hills to drive the hot air away. There is no point in taking pictures as heat-hazed shots cannot be salvaged by any imaging program. What would a picture show anyway? A desolate, parched, desert-like landscape. For a couple of hours, we follow a dusty road, and I wonder what it would be like if we had to take our lunch break in the blazing sun. I begin to suspect the sun of intending to crush me mentally and physically. Luckily, nature has a surprise in store for us as we round a corner. We find ourselves in front of an open pine forest that provides some shade and, even more importantly, offers enough dry branches and twigs for a camp fire to make food and tea – we will not have to invite ourselves into a family's house tonight!

There are some villages in this area, and so a slight chance to access a mobile network exists. When switching on my phone, this proves to be right and the signal is strong and stable – internet in the middle of a Nepalese forest! I unpack the notebook from my rucksack and, a few minutes later, am connected to the rest of the world. While I am sitting under my umbrella, surfing the net and reading and writing emails, a group of woodcutters passes by. I wonder how they perceive the situation…

Temba phones Mingma, his wife, and so we learn of a heatwave troubling the inhabitants of Kathmandu where the temperatures reach forty degrees every day. Mingma and the six-month old baby stay inside the house all day long to avoid the risk of dehydration. We are at a higher altitude than Kathmandu, but I am sure the temperature passes the 35 degree marker… exceeding liveable limits in my opinion!

The long break has done us good, but the heat is still almost unbearable when we leave in the afternoon. For hours on end a dusty and sandy road takes us over dusty and sandy hills with nerve-wracking monotony. I get angry with myself – it was me who wanted to see the holy lakes of Gosainkund and nobody else can be blamed for being parched by the heat but me. I dream of the stimulating cold of 5,000 metres, of digging my hand into the snow. Visualising the advantages of high altitude, however, intensifies my gloomy mood. Shall I unpack the iPod and listen to some energizing music? Usually, Phil Collins or Outback provide the necessary stimuli, helping me to overcome a low or to re-charge my inner 'batteries', but today I cannot even make up my mind whether I want to listen to music or not. I shuffle along the road feeling sluggish and lethargic.

Assuming the group are somewhere far ahead, I am astonished to see our luggage dumped under a tree when rounding a curve. Pimba, Sonam, Lakpa and Temba are busy collecting something from bushes that grow on the steep barren slopes on both sides of the road. On getting closer, I find orange-coloured berries hanging from the thorny twigs and after having tried one, my lousy mood evaporates. They taste delicious, and I start fantasizing about chapattis and pancakes thickly topped with orange-coloured luxury. Until today, I do not know the name of these berries, but since they resembled blackberries I made up a name for them, 'orange blackberries'. Botanists – please excuse my incompetence!

Despite the distance, the scattered houses of Okhreni greet us from the ridge, and along with a picturesque monastery with an exquisite stupa, remind me of a string of pearls with a beautiful pendant in the centre. The warm light of the late afternoon makes the colours shine. I doubt one could find a more delightful place to pitch tents. Later, when examining the workings of the prayer wheels surrounding the stupa, I cannot help smiling as the construction is a perfect example for Buddhist

Stupa with wind-driven prayer wheels

pragmatism. In order to make the wheels spin round without any further human intervention, the inventive mind of the builder came across the idea of welding together four small metal bowls and attaching them to the prayer wheels. The construction looks like one of the devices used to determine the speed of the wind and I wonder whether the tonight's prayers will fly to heaven with the aid of three or four Beaufort...

DAY 60
OKHRENI – RIPHELTOK
The unidentified enemy

The oppressive heat continues to make life miserable at low altitude, and all day long we suffer from the scorching glare of the sun. Adding to my personal woes is the presence of animals I am not particularly fond of: giant beetles whose dazzling colours could compete with the outfit of the most outlandish drag queen; huge spiders with hairy legs and caterpillars that move concertina-like through the grass. Yesterday, a vile creature stung or bit me while collecting berries. I remember the sudden pain I felt on my wrist, but thought it to be a cut caused by the thorns and so paid no attention to it.

Waking up this morning, I notice some itching on my hand and when taking a look at it, I am alarmed. The size of my hand and my forearm

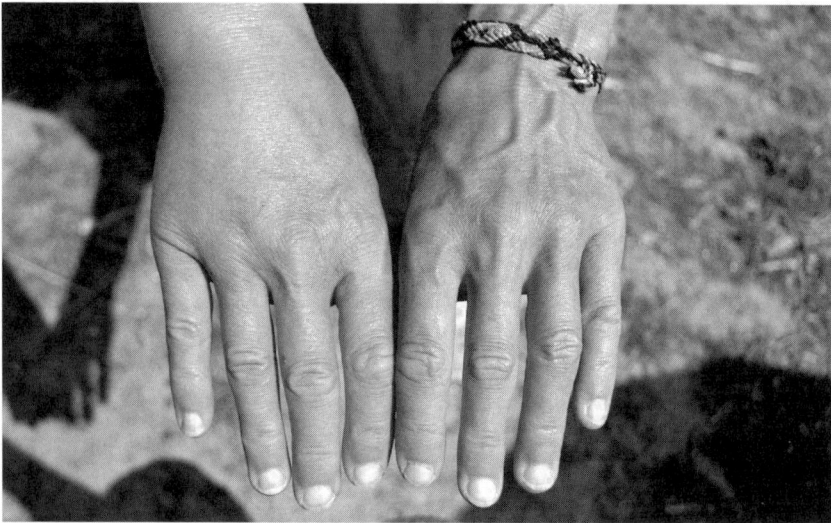

My hand does not look good

has increased by at least thirty per cent overnight and, with wrist and finger bones no longer visible, looks more like a balloon than a hand. This does not bode well. I am worried because I remember a serious infection caused by an insect bite a long time ago. Then I had waited too long before consulting a doctor and was rewarded for my stupidity with blood poisoning. Luckily, this had happened in Munich where medical help is available in virtually every street, and a doctor treated the acute infection with high dosages of antibiotics. In rural Nepal, however, there are no doctors around.

Of course, I keep broadband antibiotics for emergencies in my first-aid box, but since the hand does not show any sign of an infection, I decide to wait two or three days. If there is no improvement by then, I will join the others on their journey back to Kathmandu. I consider myself lucky because we will probably hit the next road in two days – we could have been ten days' walk from the head of a road. 'Always look on the bright side of life', I tell myself and walk off whistling the tune.

DAY 61
RIPHELTOK – SERA
They may remember me forever

Since we had an early start today, we arrive in Sera already around lunchtime. Outside the small town, we find a pretty good hotel renting out two cottages beside the river. 'This is perfect', we shout and, without hesitating, we opt for a lazy afternoon with cold beers in the attached garden.

By now, transport strikes have been going on for a while and public transport is severely affected all over the country. 'From here, we could walk back to Kathmandu in two long days', Temba says. We study the map and, while discussing different options, a bus arrives. Sonam hurries off to find out where it comes from and, more importantly, where it is heading. 'The bus will return to the capital this afternoon!' he exclaims. Everyone gets excited about the unexpected possibility of returning to the families. This is a perfect chance and, of course, they go for it. Since my swollen hand looks better, I feel no need to consult a 'proper' doctor in the capital and so will not be joining them.

Two hours later, when the bus disappears out of sight, leaving behind a suffocating cloud of exhaust fumes, we wave a last cheerful 'good-bye' to each other. We had been together for almost two months and my team's departure leaves me with a strange feeling. I miss them, and eating and drinking alone in the restaurant is only half the fun. At the same time, however, I am looking forward to continuing on my own to the holy lakes of Gosainkund.

After dinner I return to my house, groping for the switch to turn on the light… and my heart stops as I spot a large, hairy spider sitting on the wall, staring at me. Spiders terrify me and they have done so all my life.

I classify spiders in three groups:

1. Tiny ones, which I dare to squeeze under a thick piece of paper or with a shoe.
2. Middle sized ones, which I dare to attack with a Hoover hoping they have suffocated by the time I have to change the bag.
3. Large ones that give me a heart attack, paralyzing me as long as they are present. I dare not go near them!

The one in my room is a 'class 3'. No doubt about that. Mesmerized, I stare at the dark, hairy monster. What to do? Try to squeeze it? No, not a 'class 3' spider. Get a Hoover? There are no vacuum cleaners in rural Nepal. I know one thing for sure, I cannot sleep here.

I return to the restaurant, only to discover that the door is locked. Should I walk round the hotel for the rest of the night? Should I take out my headtorch and start walking to Kutumsang? Since none of these options appeals to me, I decide to knock shyly at the door. It's an embarrassing situation but, when nobody reacts, I knock louder and louder. After a while, the owner opens the door, staring at me in disbelief. From a previous conversation I know that his English is as poor as my Nepali and so I start waving around with my arms in a futile attempt to simulate a huge, slowly moving spider whilst repeating time and again, 'I cannot sleep there. I want to sleep in the restaurant'. Finally, his wife appears behind him and, since her command of English is better, she gets that something is wrong, and that my uttermost distress is related to the presence of a spider. Together, we walk to my house. Of course, the spider has disappeared, but I am convinced that the ugly, furry animal IS there, hiding under a pillow or a blanket. I can read their faces, this foreign woman is downright nuts, making such a fuss over an Insy Winsy Spider! In the end, however, they humour me and help to move all my gear to the other house. I suppose that there is also a spider here, crouching somewhere under one of the blankets, but since I cannot see it, I am able to fall asleep after a short while.

DAY 62
SERA TO KUTUMSANG
'If you don't know where you're going, but you know you've lost your way...'
(Column Sands, songwriter)

After the 'spider night' I have no problems with getting up at 4:30 and leaving the hotel half an hour later. According to Temba, it should take approximately five hours to walk over the hills to Kutumsang, following a local shortcut. Unfortunately, neither the trail nor the small villages along the route are shown on the map. Where do I go to first? I do not know – like Column Sands in his song.

The trail starts between two shops in the centre of the village and runs up to a forested ridge via endless switchbacks. By now, I no longer worry about climbing over fences – they are there to keep the hungry cattle away from the fields. I am confident enough to walk over other people's terraces, greeting them with a smile and the usual 'Namaste', just in case they object. Neither cows nor bulls with vicious-looking horns can stop me, but since rabies is said to be widespread in the Himalaya, I am careful with dogs, usually picking up some middle-sized stones whenever I hear angry barking. Nepali shortcuts are like this.

By seven o'clock, I get to a village and try to obtain more information. 'Bato Kutumsang? Way to Kutumsang?', I ask a man in a local shop where an astonishing array of products waits for the first customers of the day. 'Yes, yes… Kutumsang', he replies approvingly, nodding and smiling, and, of course, I am immensely proud of myself and walk on – and on and on; hours pass. I begin to question whether I still follow the right trail, but whenever I meet people in the fields and ask for directions their answer is, 'Yes, yes… Kutumsang', while moving an arm in a semicircle. This is not of real help to determine the direction I have to go, but I smile thankfully and walk on.

Left: Nepali shortcut through a makeshift cowshed   **Right:** In Tolkien's world

'Silly you' some will think now, 'why don't you use a GPS?' The straight-forward answer is, 'I do not know how to use one, and thick manuals frighten me'.

I hate to admit to myself that I am lost. Lost? No, never. Not me.

After some more hours, I hit a road and decide to follow it, as roads always run from A to B – let's see where this one takes me... Somehow, it is the road to Golphu Bhanjyang, only a one-hour walk away from Kutumsang. 'Well done! I knew it all along; I just wanted to see some villages', I kid myself – I tinker with the idea of buying a GPS and learning how to use it.

DAY 63
KUTUMSANG – TATEPATI
Lothlorien – Tolkien's World

When I get to the Police Check Post at five o'clock in the morning it is closed, of course. Since I do not want to wait until the police turn up, I walk on. Soon the sun will creep over the nearby hills and turn the lower valleys of Helambu into a furnace. To make sure that I do not die of thirst, I have three one-litre bottles filled with water in my rucksack. Unfortunately, this adds three kilos to my load.

Just outside the village, the well-marked trail enters a dense forest consisting of oak and rhododendron trees. Suddenly, a thick, impenetrable fog closes in and the abrupt change of light makes me feel like being carried into another world – a world void of bright colours – a sepia-toned world. The enormous trees stretch towards an invisible sky and remind me of giants from ancient times, roaming the hills and valleys. Parasitic plants, moss and lichen dangle from rotting branches create curtains, which would certainly allow the sun's rays to bring about the most impressive interplay of light and shadow. The unearthly reminds me of *Lothlorien*, Tolkien's mysterious forest from *Lord of the Rings*. Are these trees common oak trees or are they the jumbo Mallorn trees which grow in Lothlorien? Am I still in the year 2012 or did a time machine take me back to Tolkien's Years of the Trees without me noticing it? Did I enter an era when, according to Tolkien, the world got its light from two trees: *Telperion*, the Silver Tree and *Laurelin*, the Gold Tree? Then, one day lasted twelve hours, and since each of the trees gave off light for seven hours (waxing to full brightness and then slowly waning again), there was one hour of 'dawn' and one of 'dusk', where the soft gold and silver lights shone together.

Of course, I know that Tolkien did not get the ideas for his stories from visits to Nepal, but his fictional *Valinor* bears resemblance to the country, with its generally warm climate (tropical to sub-tropical), snow-covered peaks and ice in the north. Perhaps the white orchids I notice here along the trail are Telperion's silver flowers…

Admittedly, I do sometimes get carried away by my vivid imagination, but there is something magical about the forest I walk through. While sitting on a tree trunk, pondering the wonders of nature, Leia and Malcolm arrive; we met yesterday evening at the hotel in Kutumsang. They are just as impressed by the mysterious forest and its atmosphere as I am. When we resume our walk together after ten minutes, we do so with a certain respect and reverence.

DAY 64
TATEPATI TO PHEDI
Wild fantasies

A quick glance at the map tells me that the altitude difference between Tatepati (3,510 metres) and Phedi (3,630 metres) is only 100 metres. 'This is going to be one of the rare easy days', I think and set out together with my two trekking companions from Kutumsang. Locals notify us about 'a bit of up and down', but we are in a good mood and assume 'a bit' cannot be hard, wrongly, as it turns out. Soon, the trail climbs up – pretty steeply – and a steep descent follows and another even steeper ascent waits for us. As soon as my breathing and my heart rate get into a rhythm that makes me feel comfortable it has to be changed again. After a few hours, we are totally exhausted. It is like the interval training professional cyclists do for the Tour de France. We, however, do not intend to become professional racing cyclists.

The trail winds through a wild, forested region with a sense of remoteness that is enhanced by the clouds that currently shroud the hillside. Has there been a misunderstanding? Is this the right track? We are sure it is as there are green waymarking arrows everywhere – yet it was here, between Tatepati and Phedi, that James Scott, an Australian medical student, got lost in the winter 1991-2. His desperate struggle for survival lasted forty-three days before he was miraculously rescued; almost starved to death. Without food and adequate clothing, he endured freezing temperatures and excruciating hunger whilst scavenging birds circled above his deteriorating body waiting for him to die. His most horrible

Malcom and Leia

experience, however, was his growing despair over several failed attempts to locate him from a helicopter he could hear hovering above the trees.

I no longer remember who started to talk about food, but the subject keeps us busy all the way up to Phedi. The topic presumably came up after recalling the tormenting hunger of the Australian. We start with bread – something many trekkers miss while being in Nepal – and imagine eating fresh and crispy rolls with butter and delicious cheese. I go for a sixty per cent fat French Camembert whilst Malcolm opts for the goat's cheese his grandmother used to make on her farm on Malta every day. Leia is more into fruits and vegetables and talks about her Australian hometown where she used to go to the local farmers' market almost every single day to buy apples, mangos, bananas, papayas, jakfruit, pineapples, passion fruit, achachas and carambolas. Here, along the treks in Nepal, it can be extremely difficult or even impossible to obtain fruits. Our fantasies go really wild when we finally come to desserts. Ice cream topped with whipped cream, rich chocolate cakes, Crème Brûlée, Tiramisu…

And all of a sudden we stand in front of the lodge in Phedi. 'What are we going to eat tonight?' We ask each other. 'Dhal Bhat, of course', we say with a smile.

DAY 65
PHEDI – GOSAINKUND
The holy lakes of Gosainkund

The 108 lakes of Gosainkund lie at an altitude of approximately 4,400 metres and are scientifically described as oligotrophic lakes, lacking almost any algae due to nutrient deficiency. As a result, the water of the lakes is clear and of high quality. I wonder whether there are any fish – the lake would be the perfect habitat for trout.

Though the area is a famous destination for both Buddhist and Hindu pilgrims, (probably due to the number of lakes, 108, which is an auspicious number in both religions), it is mainly Hindu pilgrims who come here. Every August, during the full-moon festival of Janai Purnima, more than 25,000 devotees walk up to the lakes for a ritual purifying bath in the sacred water and circumambulate the main lake.

There are numerous legends around the creation of the main lake. A commonly accepted one derives from the Hindu mythology:

'When the God churned the primordial ocean to produce an elixir of eternal life, poison was also produced simultaneously. To save the universe from this poison, Lord Shiva drank it, scarring his throat blue in the process. To cool down his burning throat, Lord Shiva struck his trident into the rock creating three springs which, ultimately, resulted in the formation of beautiful lakes. Finally, Lord Shiva drank the water of the lakes and got rid of the pain'.

After a perfect day, which provides excellent views of snow-capped mountains and some small lakes, Leia, Malcolm and I catch the first glimpse of the main lake, situated on a ridge just off the crest. The water collected here from melting snow and ice forms the headwaters of the Trisuli River, named after Shiva's trident, the *trishula*. All along the trail, devotees have decorated auspicious places with prayer flags and set up statues depicting various Hindu deities. The black rock we can make out in the middle of the lake is said to be the head of Shiva, who saved the universe. At the end of the lake, a couple of hotels huddle together and we move into one of them, right above the water, overlooking the magnificent landscape.

The main season for pilgrims has not started yet, but there are many devotees among the guests and soon, we can make out three categories of pilgrim. There are the young, dynamic pilgrims, reasonably sporty men under the age of thirty who race from Kathmandu to Dunche on their motorbikes, run up the mountain, circumambulate the main lake and push back to the capital. There are the wealthy pilgrims. They are mostly from India, are usually overweight and often hire a horse and a horseman to help them since none of them can ride. Finally, there are the 'traditional' pilgrims. Usually older than the others, they take their pilgrimages seriously.

When we stop for food and scan the menu we are deeply shocked by the prices. Since many Indian pilgrims visiting the holy lakes of Gosainkund belong to the recently established middle classes, the locals began to charge more money for food and drink. Any tourist, complaining about astronomic prices in the Solu-Khumbu region, should come here, a glance at the menu will stop her (or him) nagging. Tonight's dinner is the most expensive in my travels across Nepal.

DAY 66
GOSAINKUND – SYABRU BESI
Jingle Bells, Jingle Bells...

Leia and Malcolm are set to continue to Dunche, so we say goodbye to each other after breakfast. I continue alone and, today, everything is perfect. The trail is wide, well-maintained and runs downhill all day long. Though the people I encounter are an intriguing mix of Hindus and Buddhists, I cannot help thinking of Christmas. A Christmas atmosphere in June? Without any snow or Christmas cookies? Yes! No matter where I am, the sound of ringing and tinkling bells reaches my ears. Of course, these chiming bells do not announce Christmas but approaching pilgrims, riding horses. The great number of devotees belonging to my category of 'wealthy pilgrim' amazes me; individual travellers (men) group travellers (single women or couples) and families. Since none of the pilgrims seems to know anything about riding, they are accompanied by horsemen who help them to climb up and down the horses, bring the animals to a halt and hurrying them along when needed. Judging from what I see, a trip to Gosainkund is good business for the owners of horses.

There's a lot of variation today, and time passes quickly. I pass through rhododendron forests, wander over vast pastures where horses and cows graze, promenade alongside wheat and corn fields and stop occasionally for tea and food. At every hotel or restaurant along the trail, the owners hand out business cards, saying as they do that 'my brother (sister, uncle, cousin...) runs a very nice hotel (or restaurant) in the next village'. Once again, the size of Nepalese families astonishes me, and so my walk turns into some kind of family visiting trip. With the shadows getting longer, I find myself in Syabru Besi and can hardly believe I have reached the small town in one day. The original plan had been to arrive here tomorrow.

Wealthy pilgrim

Outside Syabru Besi, a police officer at the check post stops me. I had passed the, still closed, check post at Kutumsang without paying and now I cannot present the receipt for which the officer asks. My experience tells me that it is never possible to predict a Nepalese police officer's reaction. A friendly looking man can be a picky bureaucrat and a wild and aggressive looking one can prove to be helpful and charming; officers are always good for a surprise. I explain my situation and make sure that he spots the Great Himalaya Trail logo on my T-shirt. Of course, he gets curious, and the matter of the permit disappears on the backburner. Before I walk on, he recommends a good hotel – belonging, of course, to his cousin.

I move into the comfortable hotel and, after an invigorating cold shower, go down to the restaurant where I meet two young Chinese students who are on a geological excursion together with a university professor from Kathmandu. Of course, I grab the opportunity to obtain more information about Chinese aid projects in Nepal and start a conversation. One of the students speaks English remarkably well, and after a while I ask why China finances so many projects in Nepal, and why China distributes free rice in the areas along the border. 'China loves to help neighbours who are in need', he explains with a broad, innocent smile. For a brief moment I am speechless, but then I touch on the subject of 'Tibet' and ask for his opinion about helping Tibet by sending tanks and heavily-armed troops. When adding that I deem the liberation of Tibet a well thought through genocide, he stiffens and his facial expression becomes hard; as hard as granite. All of a sudden, the small talk atmosphere evaporates and blatant hostility spreads across the room within seconds. The iciness in his eyes tells me that it had not been such a good idea to touch on this subject. Any attempt to return to a more relaxed conversation fails.

DAY 67
SYABRU BESI
Rest day

When I jump out of my bed, I am surprised that my left knee does not hurt after yesterday's 3,000-metre descent from Gosainkund to Syabru Besi. The unexpected absence of pain makes me think of my doctor in Munich who once saved me from ending up with a stiff knee after an operation. Even he had questioned the success of his treatment. Maybe I should send him a postcard one day?

The small town is situated in a deep, narrow valley and, therefore, it takes a while for the sun to throw her burning rays down on me, probably half an hour. Soon though, the town turns into an incinerator with the buildings and pavements reflecting the unrelenting heat. The air in my room is thick with a mixture of exhaust fumes left behind by buses and trucks and the sweet smells from a street vendor's cart where the owner prepares *Jeri* (deep-fried, pretzel-shaped loops). Children wearing school uniforms gather round his cart and wait for their turn, giggling, giving each other nudges and trying to jump the queue...

For me, it is one of the few lazy days on my four-month journey. I go through hundreds of pictures, organise them, edit a few and delete many. I update my diary and write some stories for my book – if there is a book one day. Unfortunately, my mobile internet is not working here, but I remember having seen a sign when I was here last December mentioning an internet café. It only takes a few minutes to find it again; Syabru Besi is a small town.

And what a place! Nobody pays attention when I enter the café. Four young men are sitting or lying in front of a TV watching a kung fu movie. I wait a minute or two in the hope that one of them will ask me if he can be of any help, but they continue staring at the screen and ignore my existence. Interrupting their TV session is a serious faux pas and, as a result I am treated as the nasty intruder – how dare this foreign woman disturb our peaceful life?

One of the young men finally gets up, slowly, from the bench and guides me down a dimly-lit staircase into the cellar of the building. Dirt and garbage are everywhere, the walls are stained and smeared and I expect rats or, at least mice, to make an appearance, starting a wild dance between my feet. All I find are beetles – biggish ones of course.

The room with the computers is dark, and when I ask for some light (please!) the young man switches it on reluctantly. I am baffled to see an exotic blend of dangling cables, computers, two TVs, a high shelf with all sorts of electronic equipment from the last century and at least twenty five pictures showing film stars, singers, European landscapes and some Hindu gods and goddesses. This is, without doubt, the most picturesque internet café I have ever been to.

The computer I work with seems to come from another time. It is not just slow, but extremely slow, and it takes an age to open mails and post news on my blog. Several times the machine stops working altogether and files get lost in the unfathomable depth of electronics. It is almost time for dinner when I leave the place.

Mingma, my porter for Ganesh Himal, will arrive early tomorrow morning and so, I re-organise my luggage before going to sleep.

SIX

GANESH HIMAL

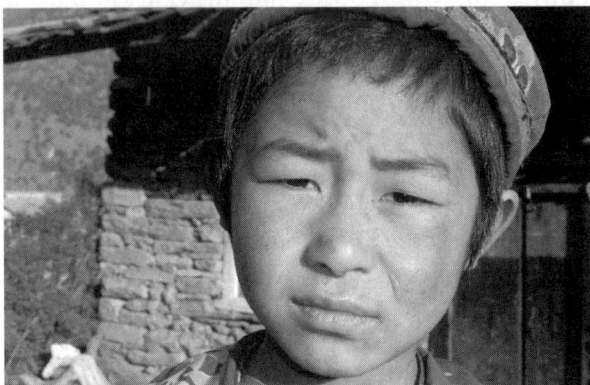

Tamang boy from Ganesh Himal

DAY 68
SYABRU BESI – GATLANG
Temptation and moral conflict

From my last journey to Ganesh Himal I recall that the trail to Gatlang is steep and exposed to the sun. There is nothing of interest along the way, neither snow-covered peaks to attract the attention of trekkers trudging up the barren slope, nor villages where one could stop for a refreshing cup of tea. Last time, I had decided to ride on a bus. 'Maybe I should get a bus ticket?' I ask myself. 'Is there anybody who would see me? Don't think so. Is there anybody who could tell others that I "jumped" a couple of kilometres? Don't think so. YES to a bus ride? NO to a bus ride?'

For the first time, I feel trapped between my desire to have an easy day and my wish to walk all the way from Taplejung to Hilsa without cheating. The moral conflict wears me out as I cannot stop thinking about a bus ticket. Of course, the heat could be used as an excuse. Everybody back home would agree with me that everything possible must be done to avoid another heatstroke. Health has to come first. To hell with stupid ambitions. The idea sounds good and makes sense, and I can already see myself sitting on the bus, being driven up the dusty road towards Gatlang.

Outdoor school in Gatlang

However, is the heat really that awful? Well… yes, it is hot, but it is not that terrible. Am I just a bit lazy? Well… yes, occasionally.

The thinking process goes on and on, and the pros and cons come and go in waves. I realise that I am stuck in a rut and despair grips me. How to get out of it? Shifting the focus helps, and so I think back to situations when I had managed well against all odds. It works and after spending some time recalling 'disasters' I had 'survived', the ambitious part of my personality is back. An altitude difference of 700 metres is not much, even with the sun trying to dry up my brain.

By the time Mingma arrives I have formulated a clear plan that sounds acceptable to me. Mingma will ride on the bus with my luggage whilst I walk up to Gatlang. Two hours later, I march off with a water bottle tugged under my arm. Soon, it happily transpires that the ascent is not as steep and uninteresting as I had feared and I actually reach the hotel in Gatlang first!

DAY 69
GATLANG – SOMDANG
Road into the future

When Mingma and I leave Gatlang, we follow the road to Somdang for a while. The original plan had been to build it all the way to Tipling or even further. Now, it is blocked by landslides, large rocks and fallen trees and it cannot be used by cars or small lorries any longer. Even so, one occasionally still meets someone on a motorbike. Soon we leave the road, winding up the mountain in a zigzag pattern, to join a trail used by the locals as a shortcut. We cross lush Alpine pastures full of flowers and walk through dense pine and rhododendron forests where white orchids grow in abundance. Just below the Khurpudanda pass, we meet a Chinese tourist and his guide and, together, we continue to Somdang – also their destination for today. The subjects of China and Tibet remain untouched.

Outside Somdang, the remains of a dilapidated power station hints at the former importance of the area. A few years back, a local metal processing company needed energy and, of course, a road for transportation. Now, the factory is shut down and the people are again without electricity or a road connecting their village with Gatlang and Syabru Besi. The locals, however, are optimistic. We run into three engineers who are busy with road-surveying work and they tell us about plans for a new power station and a road in the near future. When asked what 'near future' means they explain that environmental regulations slow down their work. 'Every single tree that has to be felled because of the new road has to be measured and registered. This takes a lot of time; probably years'.

Somdang itself consists of five or six houses and two hotels, one of which is run by three lovable sisters who have a reputation as excellent cooks. It would appear their lodge is hugely popular with Nepalese guests – the kitchen is crowded with local people on our arrival. Less than half a year ago, I spent one night here and when I enter the three women remember me at once. This does not come as a surprise since there are not many tourists travelling in this region – a shame, as a journey through Ganesh Himal is a real experience, offering insights into rural life and culture of different ethnic groups such as Tamang, Gurung, Chhetri and Brahmin, to name a few. The local people, however, do not seem interested in developing tourism and are reluctant to invest money in improving the tourist infrastructure, although, if the plans of tourism experts are to be believed, this will change.

Ganesh Himal is an easily-accessible mountain range in north-central Nepal, with the moderate elevation (below 4,000 metres) of the main trail making the area an ideal trekking destination for beginners and families. The astonishing variety of plants and animals adds to the beauty of the impressive mountain range that stretches as far as the Tibetan border. The name 'Ganesh Himal' stems from the Hindu deity Ganesha, also known as 'Elephant God'. The south face of Ganesh IV bears a resemblance to the shape of an elephant, with a ridge reminiscent of an elephant's trunk.

Tonight, I am lucky to get the only available room left at the lodge. Mingma will share the kitchen with the three charming sisters; he does not object.

DAY 70
SOMDANG – TIPLING
A woman from Tipling

Today, the minimum legal age for marriage in Nepal is twenty for men and women alike, although those aged between eighteen and twenty can marry with parental consent. Child marriage has been illegal in Nepal since 1963, but enforcing the law creates problems because legal action means prosecuting the parents. Many rural families still marry off their daughters between the ages of eleven and thirteen, usually against their wills.

Back in 2011, I met Beni, a woman from Tipling, in Kathmandu. One day she told me her story:

'At the age of twelve I was married off against my will. Still being a child this was a traumatic experience. I did not fully understand what it meant and was terrified. One year later I started working as a porter for tourists, and it was then that I began to understand the importance of education; something my parents had denied me. I packed my things and ran away'.

Today, twenty years later, she lives in Kathmandu with her two children. She has learnt to live with her past, but it is not forgotten, and all her energy goes to her work for Heed Nepal, a small organisation helping women in situations similar to hers. At the moment, she takes care of about twenty five women by providing vocational training, jobs and basic education.

Beni and I had planned to meet up in Tipling, but urgent family affairs keep her in Kathmandu and thus Mingma and I have to look for a place where we can spend the night. There are neither hotels nor lodges in Tipling, and the school where we could have camped is closed due to renovation works. The only option is to try to find a home stay with a campsite and Mingma shows that Temba was right to believe in him. His qualities and qualifications are far above those one usually expects from a porter. His English is remarkably good, he has a broad knowledge, takes on responsibilities and is self-reliant. It does not take long for him to find a place for us to stay.

The family we stay with is a classic Nepalese family comprising a couple of children, grandparents who live in the same house, and a husband who works for a trekking agency in Kathmandu and rarely comes home. It is difficult for them to make ends meet. They are autarkic farmers who grow corn, potatoes, wheat and some vegetables, but have no money left for luxuries like lentils, tea and sugar. After we have settled in, the grandmother offers a snack of boiled potatoes. I am terribly hungry and do not care that they are served with nothing else but salt and chillies. For dinner we get Dhal Bhat, which, for the first time on my journey, arrives without Dhal (lentils), and later our hosts explain that they do not have any sugar for my coffee, which I brought with me. Well, I can easily survive without sugar since I never use any in my coffee.

The fact that there is no water around on the farm, however, annoys me. After the long walk over Pangsang Bhanjyang, I would appreciate a large bowl of water to wash away the sweat and dust. I do not want to bother Mingma with additional work and so set out on my own to get water. I cross terraced fields to visit neighbouring houses; no water there. The community tap is dry as well. There must be water somewhere! But there are only empty water hoses beside the fields. Finally, when returning to our farm and asking the hosts for water, they give me a small bowl filled with valuable 'wetness'. I reckon that one of the young girls fetched it from a faraway place.

In this situation, I cannot help thinking of my work place, where we open the taps for five minutes every morning to get the warm water out of the pipes, and feel guilty about wasting drinking water whilst there are girls and women in Nepal who are forced to walk long distances with their buckets, jars and canisters.

DAY 71
TIPLING – BORAN
Even more fleas!

Washing myself and changing clothes in the family's henhouse before dinner ended in a mega-disaster for me. In the middle of the night I woke thanks to some little 'friends' in my sleeping bag. This morning I look like the top of an apple crumble pie because about 100 flea bites cover my body. Some are already filled with a sticky fluid, which makes them look like giant blisters. I get the first aid kit from my rucksack and begin to open and clean them with disinfectant. This treatment does not help against the itching, but it distracts me for a while.

I am not in a particularly good mood when breakfast arrives, and the potatoes served with salt and chilies do not help to change this. I am almost glad to leave Tipling and, for several hours, I trudge behind Mingma, sullen and grumpy.

Though the people in Ganesh Himal are Buddhists, belonging to the ethnic groups of Gurung or Tamang, I noticed on an earlier visit that the area is different to other Buddhist regions. Something was 'wrong', but it took me quite a while to put the finger on it. This region lacks all typical signs of the Buddhist religion: prayer flags flapping in the wind; walls of mani stones guiding the traveller into the village; red painted monasteries overlooking the valley and stupas containing relics of holy men. Later I learned that this area had been subject to extensive Christian missionary work, and most signs of Buddhism had been eradicated and destroyed by the local people themselves. Probably, Christian missionary work is not the right phrase because it is prohibited, and any violation of the law entails imprisonment for up to five years. Christian organisations pursue a different tactic, which is not less effective. They build schools and health posts, offer financial support, and in return, the Christian groups yield new members: 'believers'.

The population of Nepal has doubled over the last twenty five or thirty years (from fifteen to thirty million) and regions where the climate is ideal for agriculture and animal husbandry alike face serious problems

View towards Ganesh Himal

due to this population explosion. More people need more fields, more houses and more firewood, and with the rising demand, forests have disappeared. Yesterday, I experienced one of the consequences of this; a severe shortage of water.

In her book, *Der lange Abschied* (1987), the German author Dietlinde Warth describes the dangers to come and the problems the population will have to face and deal with in the future.

'The scorching heat over the barren fields creates mirages and the wind blows the dry soil away as sand. These fields lack regular water supply from the forests storing the rain… for how long will there be water flowing? For a jar filled with water, the women will have to walk for two hours down to the Markimro Khola and up the hill again'.

Dietlinde Warth's observations and worries have become reality and not only in Ganesh Himal.

DAY 72
BORAN – LAPAGAON
How far is it to Yarsa?

Last night, I didn't sleep a wink thanks to the itching all over my body. The 'anti-itch' cream is useful, but it takes time for the effect to set in and, as the enormous amount of bites means I inevitably miss a few, I itch all day long.

We leave Boran without knowing what to expect of the following two or three days. Why is it almost impossible to get reliable information about the trail to Yarsa? The inhabitants of rural Nepal do not travel just for the fun of it like me, but for two main reasons: visiting relatives and trading. Relatives rarely live far away and trips to visit them seldom improve the geographical knowledge. Trading trips, meanwhile, entail long journeys and usually result in learning the land. Unfortunately, the trade route runs to the towns in the south. Our destination lies north.

The path to Lapagaon cannot be missed and is typical for Nepal, with steep descents and even steeper ascents. It comes as an immense relief when the gradient eases before reaching Lapagaon. We walk along fields of corn and barley where people are busy pulling out weeds.

We arrive in the early afternoon and I am surprised by the size of the village. A spacious shop offers an astonishing wide variety of merchandise and a group of people haggle over prices in front of it. At the far end of the village, spicy smells wafting from a neat lodge catch our attention

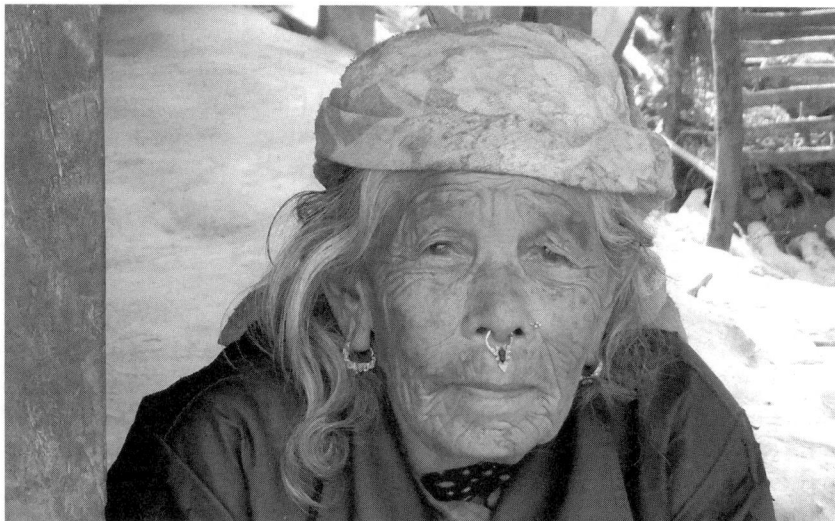

Plum lady in Lapagaon

and we decide to move in. The kitchen opens on to the dining hall and while waiting for Dhal Bhat, I get an excellent opportunity to watch the cook and his helpers busily cutting vegetables, peeling potatoes, boiling rice and baking chapattis. Apart from us, more than forty guests need to be served – most of who are restless and fidgety, rattling their plates and spoons, laughing, giggling and fooling around. Oh yes, all the guests are pupils from the local boarding school, waiting for lunch. For the owner of the hotel, it is good business that provides a regular income, independent from tourism.

Mingma walks through the village to get more information about the time we will need to reach Yarsa, whilst I sit outside the hotel, watching the people walking up and down the 'main street'. An old woman approaches me and places some freshly picked plums on the table in front of me. They are tasty, and seeing my delight, she smiles brightly and gives me some more. I try to explain that I would like to buy all her plums so that Mingma can have a share when he returns from his inquiries. With the help of the hotel owner, we soon agree on a price, and we part, both content.

Mingma returns followed by a man who is familiar with the area to the north, and he is willing to guide us to a place from where we will be able to see Yarsa. No sooner have we taken a seat to discuss some details when another man turns up beside our table and a heated discussion starts which Mingma translates. Both men claim to be experts on this region, but the information about the duration of the trip to Yarsa differs considerably; one talks about six hours, the other about two days. To be on the safe side, we decide to leave early, and I tell Mingma that, due to all the insecurities implied, he will get a double wage for the day to come.

DAY 73
LAPAGAON – YARSA
Mission Impossible I

Yesterday, we had agreed on meeting our local guide at six o'clock outside the hotel. He arrives on time and we walk off together. On our way to the next pass, Mangro Banjyang, the trail runs across Alpine meadows and passes through forested slopes and several times we meet herders driving their animals to fertile grazing grounds. None of them seems to be in a hurry, and I am thankful for every break we take with people along the 1,200-metre ascent. I am tired and exhausted. On reaching the pass we lunch together with a group of shepherds on their way to one of the makeshift huts further down. While their goats pluck off all the leaves and twigs they can reach, we eat chapattis, boiled eggs and plums. How lucky they are, in a way. To them it does not make any difference whether they arrive at their shelter in one hour or in three. We, meanwhile, still have a long way to go, and we still do not know how long it will take us to get to Yarsa. Obviously, our guide knows what I am thinking and brings me back to reality, 'Let's go!' he says without mercy.

The trail contours the ridge, and shortly afterwards we cross another minor pass from where we finally start climbing down into the main valley. Up to now, trail finding has not posed any problems, and Mingma and I begin to wonder why it is recommended that tourists find a local guide for this section. Is it just a trick to provide income for the locals? As we proceed with our descent, we get the answer. There are trails everywhere, running in all directions, but none looks like a major path. Which one to choose? Now, we are happy about having the guide with us. On reaching a shepherd's hut, I identify a well-used track and head for it, wanting to demonstrate my inner (female) intuition and sense of direction. 'Stop! This path just runs to the next village, and from there you cannot walk on; dead end street', our guide explains. We follow him over lush, green, grazing areas, passing another hut before finally coming to a place from where we can see Yarsa on the other side of the valley. It seems almost close enough to touch.

Mingma and I are optimistic and believe that the walk to Yarsa will take less than two hours. We are soon proved wrong. It turns out to be impossible to climb down to the Richel Khola and cross it as the valley is too steep. The trail, now marked with green arrows (to signify hope?) leads down and down and down; the distance between Yarsa and us increasing with every single minute. Maybe the guy who estimated two days had been right...

In the late afternoon, we arrive at a new bridge where a group of porters are taking a breather before crossing the river and climbing up the zigzagging trail to the high plateau of Yarsa. My courage evaporates when looking at the hundreds of switchbacks rising behind the bridge. I am tired, my feet are tired, and my shoulders hurt. It does not help wallowing in self-pity. 'Your enemy is apathy', Deepak Chopra once wrote. Correct, and then I remember another sentence 'You are able to do or to achieve whatever you want, but you have to want it with all your heart and energy'.

Tonight, I want a good meal and a decent place to sleep.

DAY 74
YARSA – KHORLABESI
Mobile shops

We found a good dinner and a perfect place to stay, of course. Yarsa had come into view just before the darkness of the night fell, erasing the colours painted by the evening sun and enveloping the country in the dark nothingness of a moonless night.

Our breakfast consists of numerous Tibetan breads with jam and fried eggs, and we consume everything greedily under the watchful eyes of the villagers who have come to pay us a visit on the balcony. My coffee arouses curiosity, but when I ask them to join me for a cup, only one of the locals wants to try it. The others prefer tea.

A glance at the map tells us that the ramble down the hill to Machha Khola will be short and easy and so we set off in high spirits. Rounding a corner half an hour later, we bump into the group of porters we met yesterday at the bridge. Their overloaded dokos are filled with a wide range of paraphernalia: cups, cooking pots, metal bowls in different sizes, plastic buckets, several rolls of fabric, tools, jars and plates – and I reckon that there are more treasures hidden deep down in the wicker baskets. By now, my Nepali is sufficient to ask and to answer basic questions, but not good enough for sophisticated conversations; Mingma has to translate.

I learn that these porters are self-employed and on the way to remote villages and farms in this area where they will try to sell their goods, in a similar fashion to the hawkers or pedlars who walked to faraway settlements in the Alps. This form of work guarantees freedom from greedy and poorly-paying owners of shops and hotels, but the job is backbreaking. Inevitably, my thoughts drift to Nabin and Lakpa, the two young brothers I met in the Kangchenjunga area. Will they ever have a chance to save enough to buy their own stuff and work as independent hawkers? Or, will they remain trapped in a system of exploitation and abuse for the rest of their lives?

During the civil war, the Maoists began to fight the exploitation of porters by agencies and trekking groups – a praiseworthy objective. Along the main routes, the Maoists interrogated trekking porters about how much they earned and how much they were requested to carry. If they decided that porters were asked to bear too much the Nepali Sirdar, or trek leader, was forced to take a particularly heavy load himself, and the 'donation' they asked for from the tour leader and the tourists ranged from a few hundred rupees to 10,000 rupees per foreigner; approximately 100 Euro.

Since the foundation of the International Porter Protection Group (IPPG) in 1997, hundreds of volunteers have worked hard to improve the conditions of mountain porters in the tourism industry worldwide. One of their goals is that: 'No porter should be asked to carry a load that is too heavy for their physical abilities (maximum: thirty kilograms in Nepal).

Mobile shop

Weight limits may need to be adjusted for altitude, trail and weather conditions; experience is needed to make this decision. Child porters should not be employed'.

Even given many improvements that could be brought about, the conditions many porters in Nepal work under are still some way off the IPPG's recommendations. Whenever and wherever I walked in Nepal I met porters employed by renowned foreign companies carrying forty kilograms or more. Some of them were under sixteen years old. Working for local employees and delivering goods to hotels and shops can be even worse. Many porters freight loads exceeding their own body weight and the negative impact on their health is alarming: deformation of joints, arthrosis, lumbar disc damage, scoliosis and reduced life expectancy to name a few. Who looks after the rights of these porters?

The hawkers carry on to a village up the hill; we follow the pleasant trail down to Machha Khola. On arriving at the village, we stop for lunch and to sort out some of the gear. Mingma will continue his journey south to catch a bus back to Kathmandu and I will walk to Khorlabesi, in the north, to meet Temba.

On the way, I meet a group of horsemen who stop me. 'Your guide Temba is already waiting for you in Khorlabesi', they tell me. I had not expected anything else.

SEVEN

MANASLU

Mt Manaslu

DAY 75
KHORLABESI TO PHILLIM
Discrimination of tourists

Temba and I make the most of the invigorating fresh air of the early morning, leaving Khorlabesi at six o'clock without breakfast. Regarding the landscape, this stage is certainly the most impressive along the Manaslu Circuit. Not far from the village we pass hot springs, which would be perfect for taking a bath if we had time to linger. After crossing the Budhi Gandaki, with its water swirls and cascades, we wander through an impressive forest before arriving in Dobhan where we have breakfast with a small group of engineers who are on their way to Samdo.

Now, the scenery changes dramatically. The valley narrows, and the trail runs above the deep gorge that the Budhi Gandaki has cut into the rock. The cliffs rising on both sides are almost vertical. The crystal-clear green water seems to boil and the deafening sound of the river careering down the narrow canyon echoes through the forest like thunder. For hours on end, we are accompanied by numerous impressive cascades, torrents and rapids.

When we stop for lunch and order Dhal Bhat in Jagat, I experience discrimination of tourists for the first time. Whereas the owner of the restaurant serves Temba a capacious plate with an Everest-sized heap of rice, I only get a small plateful (although it too, is a perfectly-shaped mound).

'What's that?' I ask. 'Why does Temba get more than me?' The answer comes quickly, 'He is Nepali and tourists always eat less'. 'Nonsense', I reply, 'I want as much Dhal Bhat as my guide. That's discrimination of tourists'.

Smilingly the owner returns to his kitchen and returns with a generous second helping for me. I assume his plan is to make me give up; poor man. He does not know that I have been walking for more than seventy days and capable of winning any Dhal Bhat eating competition. He stares at me in disbelief as all the rice, curry, pickle, and lentil soup disappears from the plate and ends up in my stomach. Temba and I can hardly stop laughing as we wave good-bye to the owner of the hotel and leave him standing there, completely puzzled.

From Jagat, the track climbs up a terraced hill from where we enjoy a stunning view of the Sringi Himal. Less than two hours later, we arrive in Phillim and at the hotel we meet the engineers again. On the one hand, their planned itinerary impresses me and confronts me with limits imposed upon me by age. On the other hand, two of them already complain about headache and breathlessness. Age entails certain advantages since being forced to walk slowly reduces the probability of high altitude sickness.

DAY 76
PHILLIM – PROK
Deceptive adverts

On the way along the Manaslu Circuit, a colourful sign praising white sheets, clean toilets and good food attracts my attention. According to the given information, this auspicious place is in a village called Prok which is off the main tourist route. Needless to say, I am always tempted by good food and clean toilets and, thus, Temba and I decide to spend the night there.

Locals tell us about a shortcut to the village, which lies a thirty-minute walk away from the main trail. Shortcut? Maybe it is the shortest route but for sure not the easiest or most comfortable one. Following the short cut entails climbing up a hill (400 metres altitude difference), crossing a jungle-like forest and then walking down the hill again before finally plodding up a sandy zigzagging track to Prok. In my opinion, a clean lodge is worth taking on some hardship.

The houses of the village are widely scattered over the plateau and we do not know where to look for the hotel. We ask everyone we meet for directions, but the answer we get is always the same: 'Thakali Lodge? Sorry, no idea. Maybe this way?' and is accompanied by an indecisive wave of a hand to the right or to the left, before they return to their work in the fields. I am tired, frustrated and angry with myself and that is never a good sign because it leads to nothing but more irritation. We find two hotels in Prok, but the standard is poor, and the mere thought of having to stay in one of them for the night makes my skin itch violently.

After a long and fruitless search in the widespread village, we bump into a man who has the key to a new, but still not opened, hotel and, after some negotiation, we are allowed to stay there for the night. This sounds promising because a brand new hotel will provide a high standard, but then we learn that nobody works there; the kitchen is closed. We have neither food nor any cooking equipment and going to bed with an empty stomach does not appeal to me, at all. The man seems to notice my despair and offers us a meal at his house. We accept thankfully.

The darkness of the night has already arrived when we walk back to the hotel that lies at the far end of the village. The maze of trails running through the fields makes orientation difficult and, a few times, we have to think twice before opting for one of the tracks at a junction. The high plateau is bathed in a pale, ghostly moonlight, and all of a sudden the quietness of the village seems almost sinister. I try to keep up with Temba walking in front of me, the light from his head torch jumping up and down. When we get to the hotel, I am surprised to find it transformed from an inviting home (daylight) into an ominous looking pile of stones (night). The hallway is pitch-dark and while groping my way along the lightless, empty corridors to the room, I cannot help thinking of the Overlook Hotel from Stanley Kubrick's movie *The Shining*. Thankfully, however, this is a brand new hotel without 'memories', and Prok is hardly the place for Jack Nicholson to spend his holidays. I hope.

DAY 77
PROK – SHO
Mr Bean

Jack Nicholson did not pay us a visit last night and, therefore, everything was quiet and safe. Temba locks the main entrance of the hotel before leaving for breakfast in the village. Our host's house resembles a terraced

Terraced house in Prok

house with not just one family living in the long stretched building but four and the private rooms on the second floor are only accessible via a ladder. The basement houses animals and supplies.

The day starts with an easy stroll down a scree trail back to the bottom of the valley where we meet the main track again. After a while, the valley narrows, and we walk through a broad-leaf forest before arriving at a spectacular canyon carved out by the river. It is a wild and unspoiled land, waiting to be explored by the visitor. In his guidebook *Great Himalaya Trail*, Robin Boustead terms this area a hidden gem which is, indeed, no exaggeration. All those who want to experience authenticity without internet and hot showers should come here. As for landscape, the Manaslu Circuit can easily compete with the Annapurna Circuit Trail and is, in my opinion, even better because you are that much nearer to an 8,000-metre peak. Temba and I saw a group of porters yesterday but no tourists and so I can easily convince myself to be the only trekker to explore this area.

We had reckoned to reach Lho today, but on arriving in Sho, it starts raining, and stopping here for the night seems to be a sensible decision but we have hardly ordered tea when the sun is back. However, since I am mentally no longer prepared to continue walking, we settle into the hotel. It is a cosy place with clean, spacious rooms, a comfortable dining hall, a well-stocked shop offering all sorts of little luxuries for trekkers, and a bathroom with running hot water delivered in a big bucket.

The owner of the lodge tells us about a group of young engineers who had stopped for tea yesterday afternoon, but today returned from Lho and walked back down the valley; two of their number suffering from high altitude sickness. We probably missed them when eating lunch.

After a bucket shower, I feel fresh and set out to explore the neighbourhood. A young Sherpani is sitting in a front garden, weaving multi-coloured pieces of cloth which she will stitch together later to make the traditional apron that Sherpa women wear. The handmade loom is basic but seems to function perfectly. Her fingers work the homespun woollen threads with somnambulistic accuracy while her feet operate the pedals, controlling the mechanism of the loom in the way a musician operates the pedals of an organ. I ask for her permission to take some pictures and a video; smilingly she consents. Later, we take a look at them on my computer, accompanied by laughter.

Returning to the hotel, I find the dining hall full of Nepalese people sitting in front of the TV. There is no satellite TV in Sho, but the hotel owner has a DVD player! In a situation like this, I would normally expect a Bollywood movie or a Chinese kung fu movie to be presented to the audience, either in Hindi or Chinese. It is not always that I fancy watching films in languages I do not know, and I am already on my way to the room when I find out that there is a special programme tonight: *Mr Bean*.

I have to admit to belonging to the group of people who love Mr Bean, and I remember times when six o'clock on Monday evening was a sacred time – Mr Bean time! Also the people in Lho seem to love him. When the movie starts, a strange mix of spectators is watching *Mr Bean's Holiday*: a few porters who, probably, have never attended school, teachers from the village, a couple of rough-looking horsemen, Temba and me. We clap hands and laugh until tears roll down our cheeks, cheerfully and happy. Words are unnecessary to find a common denominator for joy and excitement.

DAY 78
SHO – SAMDO
Protecting the locals

Behind our hotel, the majestic Mount Manaslu, also known as *Kutang*, greets us. The name means 'mountain of the spirit' and derives from the Sanskrit word *Manasa*, meaning 'intellect' or 'soul'. Mount Manaslu, the eighth highest mountain in the world, was first climbed on May 9th, 1956 by members of a Japanese expedition.

We follow the ancient salt trading route that originally connected this area with Tibet and played a vital role in the economic situation of the local villagers. Today, this route makes an excellent trail for trekking tourists on their way round the mountain and tourism provides additional income to farming and animal husbandry. One hour after our start from Sho, we reach the village of Lho, which is dominated by a large Gompa, including a boarding school for more than 150 children. I stand below the impressive monastery, which towers on a hill above me. Set against the backdrop of the mighty snow-covered Manaslu, it is an awe-inspiring place. From here one has, undeniably, the best view of the mountain with its two peaks. This valley is a haven for many highly endangered animals like Snow Leopards, Pandas, Himalayan Tahr and Musk Deer because monks from the monasteries in the area put a hunting ban in place and thus help the wildlife to thrive.

Until today, I considered the monastery in Teng Boche, in the heart of the Everest region, the most beautiful place in the Nepalese Himalaya. Now I have a new favourite; the Ribum monastery in Lho. This location offers a breathtaking view towards Manaslu as it rises into the sky just behind the monastery. It almost seems as though these peaks look straight into the monks' rooms. A perfect place to stay for a few days: to rest, to wonder, to explore and to find peace and relaxation and to regain one's strength before continuing the trip over the Larkye La.

Unfortunately, I have a tough time schedule to follow, and so we move on along fields of barley and buckwheat, through pine and spruce forests until we arrive in Sama for lunch.

Monastery in Lho

The village is the starting point for expeditions heading for Mount Manaslu which, together with tourism, play a pivotal role in the local economy. Expeditions are usually planned abroad, but they are implemented and conducted with the help of Nepalese agencies who procure the necessary permits and put together the support groups for the climbers. Though these agencies rarely employ guides or porters on a regular or permanent basis, they have a 'pool' of people they work with and who normally live in Kathmandu. When the expedition groups leave the capital they not only take the entire team from there, but also the food. What is left for the people living along the approach route? Probably just waste. The inhabitants of Sama have introduced a job-creation scheme to improve their financial situation, only locals are allowed to work for expedition groups beyond the village. This regulation entails enormous logistic problems, but in my opinion, fighting poverty of local communities has to be prioritized.

Samdo, our destination for today, is the highest village in the valley of the Budhi Gandaki. The villagers, who are of Tibetan origin, were ceded the land by the King of Jumla over 500 years ago, but only claimed their property after the Chinese takeover in the early 1950s. Then, they came across the border from the village of Riu in Tibet and built a new village here, at the location of their old herding settlement. Since then, they have established a trade with China and India, marketing among other things, Yartsa Gunbu, that grows in the region.

On our arrival, the settlement seems to be utterly abandoned; locked doors, no smoke from the chimneys, not a single soul to be seen… We knock on doors and shout, but nobody answers. Will we have to sleep in the open? I do not like the idea. It has started raining, and an ice-cold wind blows through the main street of Samdo. Temba and I find shelter under a roof and try to get as comfortable as possible on a bench, waiting and hoping for someone to pop up. I do not feel well. My throat is sore and fever chills set in.

Of course, some locals must have seen us and informed the owner of the hotel of the two people sitting in front of his house because, all of a sudden, he stands beside us, promising us a meal that will be ready in one hour. I appreciate promises like that. What I do not like is the news he comes with, the keeper of the lodge below the pass where we intend to stay tomorrow has left for Kathmandu. The hut is closed. Well, this means we have to walk a two-day trip in one go – including crossing a 5,000-metre pass.

DAY 79
SAMDO – BIMTANG
Bad news and Mission Impossible II

Yesterday evening, news of the closed hut played on my mind and when I wake up to an even more sore throat than yesterday, I think about giving up. It is a 1,400-metre climb from Samdo up to the Larkya La, and even fit and well-acclimatized trekkers need about ten hours to reach Bimtang, on the other side of the pass. Most trekkers walk this bit in two days.

I do not know what to do, but after some back and forth thinking, I get an idea that sounds reasonable to me and discuss my plan with Temba, 'Ok', I decide, 'Let's give it a try, but if we haven't reached the hut before eleven o'clock, it will show that I am too weak for the crossing in one day, so we'll return to Samdo'. Temba probably thinks I am totally crazy, but he nods, and ten minutes later we set out.

We are not the only early birds today. Wherever we look, there are locals crouching on the slopes, staring at the soil. What are they looking for? Of course, I have already spotted all the flowers growing in abundance: primulas, iris, saxifrages, edelweiss and many more. No, Samdo's inhabitants are not here to pick flowers but to look for that 'Nepalese gold', Yartsa Gunbu. This explains why the village was virtually deserted on our arrival.

Just before my self-established deadline expires the hut comes into sight. Is this a good or a bad sign? If I was honest with myself I would admit it to be a bad sign, normally, I would have reached this shelter by ten o'clock. Of course, I am successful in telling myself that meeting the deadline is the critical component in the calculation and so all worries are suppressed – including the fact that I feel sick. After a short break, we continue our ascent.

The higher we get the more the landscape turns into a stony desert. In my opinion, the Larkye La is the ugliest pass I have ever crossed in my life; just stones, sand, pebbles and rocks. This would be an ideal location to shoot a film about a moon landing. The views from the highest point are said to be wonderful: a mountain panorama comprising Himlung Himal, Cheo Himal, Kangguru and the towering Annapurna II. All I see from the highest point are clouds in different shades of grey.

We eat our lunch crouching between the remains of a stone shelter that protects us from the ghastly cold wind that blows from all directions. Neither the cold nor the inhospitable surroundings makes us want to stay longer than necessary and so we attach some Tibetan prayer flags to an old rope hanging down from a wooden pole, yell 'Ki ki so so lha

On the way to Larkya La

gyalo' (may the gods be victorious) and begin the long descent to Bimtang.

The zigzagging path is steep and energy-consuming. Desperately, my eyes scan the landscape in the faint hope of spotting the village; no luck. The trail seems to run down forever, ending at the gate of eternity. Then, all of a sudden, the gradient eases and after crossing a few moraine hills, Bimtang is right in front of us. After ten hours, a 1,400-metre ascent and a 1,500-metre descent, I am hungry and long for a bed and a rest.

The owner of the lodge tells us about a trekker who had come over the Larkye La five days before, on his own. This man had ordered two plates of spaghetti and eaten the lot. It was probably Nicolas, the guy from Switzerland.

DAY 80
BIMTANG – DHAREPANI
For my liking – a perfect day

It is a chilly but beautiful morning as the sun hits the peaks around the valley. After having slept for ten hours, my body feels reborn and I can hardly wait for breakfast to be served. Several Tibetan breads disappear from the plate, followed by a giant omelette and an even bigger chapatti. This will do until we stop for lunch, I hope.

In Bimtang the construction business is in full swing and, despite the early hour, the sounds of hammers, saws and slicers already can be heard all over the place. Just beside our hotel, people are busy building cottages reminiscent of the small huts one finds on Norwegian camp sites. I envy the tourists who will come here next year, who will have a magnificent view of Manaslu and Larkya Peak straight from their new, cosy accommodation. Apart from the hammering and sawing, I can only hear the soft whistling wind and occasional cries from crows circling above the nearby Tibetan cemetery.

During the main season, about ten people live in the quietness of Bimtang, but in the winter only one person stays in this remote place: Ryalbo Lama, a Tibetan refugee. He enjoys living here, surrounded by the emptiness of a dramatic environment, and leaves the village only to do some shopping in Kathmandu three times a year; a tough ten-day round trip from his house.

We leave the rich grazing fields of Bimtang, stroll down the hills through open pine and rhododendron forests, and cross dense jungle areas full of giant ferns and orchids. At our lunch stop, we get the best Dhal Bhat we have ever eaten along the trek, accompanied by spicy mushroom curry. This meal will certainly be top the chart list for a long time.

Eventually, after a long but enjoyable day, we get to the large village of Tilje, where the inhabitants are a mix of Manangis (Tibetan origin) and Chhetris (Hindus). Both the architecture and the menus in the restaurants reflect the ethnic variety. We spot flat-roofed stone and mud houses, typical for Tibetans, and the red and white painted homes of the Chhetris. Food wise, one finds rice dishes and sweet tea with milk in the restaurants run by Hindus, while one can order Tibetan Tea and Tsampa in the lodges owned by Manangis. Unfortunately, we do not have time to fully experience the ethnic diversity as we want to push on to Dharepani today.

When we finally arrive the sun has disappeared behind the steep walls of the Marsyangdis valley, a valley marking the land of apple pies, cold beers, hot showers and prayer flags fluttering in the wind. We have reached the Annapurna Circuit.

Left: Building new cottages in Bimtang   **Right:** Hotel on the 'Apple pie Trek'

EIGHT

ANNAPURNA/UPPER MUSTANG

Annapurna massif

DAY 81
DHAREPANI – CHAME
The future of the Round Annapurna Trail?

Today it is time to say good-bye to Temba – from here, I will follow the Round Annapurna Trail on my own and Temba will return to Kathmandu to get a few things done. He has to put together a group of porters for Dolpo and Mugu, apply for the required permits, do the necessary shopping and, of course, he will have a few days together with his family. One more time, we work our way through the long to-do list and after a late breakfast, we set off from Dharepani in different directions.

The last houses of the village lie behind me when I meet a small group of mountain bikers. This, as such, does not come as a surprise since the trail around the Annapurna massif has developed into a 'must' for every enthusiastic cyclist. Even as far back as the '80s, locals talked about people taking their bikes on the plane to Humde to cycle along Manang valley.

The Round Annapurna Trail is considered one of the most popular trekking routes in the world and in 2011 approximately 90,000 tourists visited the region. Unfortunately, the legendary trail will soon be history.

In 2005, the Nepalese government started an extensive road-construction project in the Annapurna area. One road would run from

Road construction along the Annapurna Circuit

Besisahar to Manang and one from Beni (near Pokhara) to Lo Manthang in Upper Mustang. The Chinese had already built a dirt road from Tibet to Lo Manthang and the extension to Pokhara was intended to stimulate development. That was the plan. It will be some time before tourists can travel from Besisahar to Manang by bus or jeep, but the road between Pokhara and Jomoson was completed years ago, and the trip now takes seven hours instead of seven days. Trekkers who wish to travel on to Muktinath or Lo Manthang can do so by jeep, despite a pedestrian-only suspension bridge at Jomoson – on the other side of the river, more jeeps wait to ferry them onwards. There is no doubt that the end of the Annapurna Circuit is close – just imagine walking along a dirt road and being overtaken by a 4x4 full of tourists toting bulky cameras...

What does the road mean for the locals? There are both advantages and disadvantages. On the one hand, it means better infrastructure, facilitated access to health services and schools, and reduced costs for transportation of goods. On the other, many tourists no longer stop and spend money in the small villages along the trail, porters are without jobs, and there are empty hotels and restaurants. There are plans to build a new trail, but the when and where remains unclear.

It is difficult, if not impossible, to envision future developments. The first hotels in the Khali Gandaki valley have closed for good, whereas locals still build new hotels in Manang Valley. For sure, other trekking areas in this region will benefit from the road as it provides facilitated access. The 'winners' will probably be Nar/Pho, Manaslu and Lamjung Himal.

Maybe there will be opportunities we do not see right now. The government talked about plans to introduce a different kind of tourism, based around 'adventure' travel rather than the basic tea-house trekking. It may, eventually, be possible for tourists to drive through the valley, hop out to take photos and travel on.

On my way to Chame, I try to avoid walking on the road where construction work is in full swing. Whenever possible I choose trails used by the locals as shortcuts, but wherever I am, the sound of heavy machines reverberates through the valley.

DAY 82
CHAME – LOWER PISANG
Safe drinking water stations

As long as I travel together with my team there are usually no problems finding safe drinking water, but walking on my own can be difficult now and then. Previously, a pocket water filter saved me from encountering problems that make other tourists scurry for the toilet. On this long trip, however, weight is one of the key factors and so I left my water filter at home.

All over Nepal, the quality of water is extremely poor and, therefore, many tourists rely on drinking mineral water which is widely available along the major trekking routes. On the one side, this is sensible, but on the other side, the plastic bottles are extremely harmful for the environment as solid waste management is a big issue in Nepal. There are few, if any, recycling units in rural areas and as a result huge heaps of discarded plastic bottles are virtually everywhere: beside roads, behind houses, under bridges, in forests, between rocks, beside rivers and in creeks.

In the villages along the Round Annapurna Trail, the local communities work hard to solve this problem. In 2000, the New Zealand government sponsored the establishment of ACAP safe drinking water stations around the Annapurna circuit. ACAP is responsible for training technicians, and local women's groups maintain and manage the stations in the villages; using profits for community projects. Today, there are sixteen sales points along the Annapurna Circuit. The purification systems use ozonation to produce water at lower cost than those bottles carried in by mules or porters, and thus trekkers can purchase safe drinking water at a fair price. If all travellers bought water from these stations, waste could be reduced by as much as 100,000 plastic water bottles, each year! An additional benefit of these sales points, which I am very happy about, is that being able to refill my water bottle in virtually every village saves me carrying an extra three kilos on my back.

Before leaving Chame, I visit the local primary school and have a long talk with the old headmaster and some of the teachers. There is much to say about autism, but at the same time, I also want to look at a project that Mountain People, an NGO I am involved in, paid for last autumn. Previously, the playground could not be used because it was too steeply angled. Now, a supporting wall runs along the school yard, which has been levelled, and there are new white boards in the classrooms. Proudly, one of the teachers offers me a sightseeing tour through the building and invites me to a private performance into the school yard.

The pupils stand, soldier-like, in front of the headmaster who rhythmically hammers with a long wooden stick. He seems to be a highly respected and feared person since none of the children even moves the little finger. At a signal, they start reciting the English alphabet at top volume, followed by two songs – also at top volume. It is difficult to decide whether I want to break out in laughter or tears because this demonstration reminds me of my days at school when drill was the only accepted teaching method. After the performance of Nepalese pedagogy, I meet a teacher who represents a new and modern way of thinking. Over a period of two years, he has spent most of his holidays collecting information about the different local cultures; songs, dances, customs and religions. In a year or two, he will present his local curriculum to the government. This will certainly lead to heated discussions because local peculiarities are generally ignored at Nepalese schools.

The stage between Chame and Pisang is a varied trip. There are several small villages, forests with pine trees and spruces, copses of birch trees and rhododendrons and pastures with thousands of flowers. In the middle of a forest, I bump into a couple of journalists from Pokhara, who have joined an 'anti-rubbish' group and are returning from a 'cleaning job' in the region of Nar/Pho. Of course, my *Autism Care Nepal* T-shirt, with the additional logo of the Great Himalaya Trail instils curiosity, and they interview me on the spot. 'Publicity is always good and worth a delay', I think when walking on half an hour later.

Just before arriving in Pisang, I have to decide whether to continue my walk along the bottom of the valley or to follow a trail running above it. Since I know the high route from previous trips, I opt for the easy alternative and look for a nice hotel in Lower Pisang.

DAY 83
PISANG – MANANG
Watching movies

The Annapurna Circuit evokes many memories for me – I have been there a couple of times since the '80s, visiting in all four seasons. From previous trips, I recall lodge keepers handing out candles in the evening so that guests would have some light in their rooms, and I remember almost empty shops in the rainy season, when it was actually impossible to get even a package of biscuits, let alone luxuries like honey or jam. Now, the small shops along the trail offer all kinds of products to make a trekker's heart leap.

On my way to Manang, I visit another school project the organisation Mountain People supports. Only half a year ago, the school received the urgently needed financial means to convert two unused classrooms into a kindergarten. I am pleased to see that all the work has been done as this is not the norm, but the locals were genuinely interested in making a change. We financed the material and they did the work of painting the walls and covering the dirt floor with wooden planks. The meeting with the teachers is accompanied by several cups of tea, and during this informal togetherness I grab the opportunity to hold a short lecture on autism. The teachers show strong interest in a topic none of them has ever heard about before.

Manang, situated at an altitude of about 3,500 metres, is the biggest village in the district and my final destination for today. The inhabitants are popularly known as Manangi and are said to have emigrated from Tibet

Old houses in Manang

several centuries ago. Traditionally, they have never been solely farmers, but also keen traders and, due to a special dispensation from the king, they were allowed to do business in faraway South East Asian countries. Some Manangis would stay away from home for as long as six months per year. The most important trading goods were precious stones, metals, musk, herbs and other items. Today, the people here still live on trade, animal husbandry and agriculture, but since the opening of the valley for foreigners in the '70s, tourism has become the main seasonal business.

Now, there are not only about 150 lodges in this region but also a few internet cafés, a visitor's information centre, a cultural museum and a health post. The staff of the latter provide general medical help for locals and tourists alike. They hold lectures about high altitude sickness and conduct rescues, for which they are perfectly equipped, with bottled oxygen and a Gamow Bag (portable plastic hyperbaric chamber).

Strolling through downtown Manang, I come across another sign of modernity, which had not been there on my last visit twelve years ago: the Projector Hall, a local cinema. I am known for being a passionate movie-goer and, with some Australian trekking friends whom I met in Pisang, decide to pay it a visit later.

The cinema is a dark, spacious Spartan room in the basement of a local shop. Instead of the plush covered comfortable armchairs the audience is used to in Europe, we take a seat on rickety, wooden benches with fluffy yak skins spread over them. What an atmosphere! This is the perfect setting to watch films like *Caravan*, *Into Thin Air* or *Seven Years in Tibet*. Since we are the only spectators tonight, we can even choose the film. Movies on demand! We opt for *Seven Years in Tibet*. It is not long before the owner offers tea and popcorn – which are included in the two Euro ticket price! Our plan for tomorrow night is… ?

Left: Cinema in Manang **Right:** Sandstone formations near Manang

DAY 84
MANANG
Rest day

My last rest day was two weeks ago and so I decide to have a lazy day: writing and answering emails, eating cakes, visiting Braga Gompa and wandering around to look for a specific prayer wheel.

In 2000, I visited Manang with Uli and whilst strolling through the medieval lanes, discovered an unusual prayer wheel, a brand new, shiny and colourful Nescafé box, probably integrated into the long line of traditional prayer wheels by a modern (?) devotee. Is it still there?

Manang has changed a lot since then, and I am about to abandon my search for the prayer wheel, assuming that it only enjoyed a short life, when I find it again. The colours have faded, and the once shiny surface is dull and brownish, but when I push it, it still spins the same way it did twelve years ago.

I walk on to the Tibetan-style settlement of Braga where an old Buddhist monastery is perched on a high crag overlooking the village. The monastic building bears a resemblance to a bird's nest that is attached to the rock and which only can be approached via a steep path. I follow the zigzagging trail winding between partly derelict houses that seem to be built one atop the other. Each one has a spacious veranda formed by the neighbour's rooftop. Previously, monks inhabited the buildings, and the size of the site allows conclusions to be drawn about the former importance of the monastery. Now, the houses seem abandoned, and nobody walks the narrow lanes. Arriving at the main door, I am disappointed to find it shut with a heavy, rusty padlock dangling down from a chain. In the slight hope of bumping into a local, I round the building, but there is nobody, and so I decide to walk back to Manang. When scrambling down the zigzagging trail, an old woman appears between the houses; shouting something in Nepali to me while rattling with a bunch of keys. She is some kind of janitor holding the keys to the monastery.

In many Buddhist monasteries, visitors are not allowed to take pictures inside, and so I am surprised to learn that this rule does not apply here. Luckily, I had recharged the camera batteries at the hotel yesterday because hundreds of fascinating objects draw my attention.

Threatening faces of demons stare at me, masks for religious festivals hang on the walls, and exotic musical instruments are lined up on shelves. I am fascinated by the colourful paintings that cover every single square centimetre of the walls, the ceiling and the supporting pillars.

While taking pictures in the prayer room, the old woman places her offerings in front of the altar, starts reciting mantras and lights incense sticks and butter lamps. All of a sudden, I feel like an intruder who disturbs the deep religiosity of this woman. Even after having been to Buddhist countries many times, my knowledge about this religion is appallingly limited. This, however, has nothing to do with Buddhism as such, but is a result of a general lack of interest in religions.

Although Buddha's teachings are atheistic, the people in the Himalaya do not deny the existence of god-like beings. They consider the universe a place populated by a great number of celestial Buddhas and Bodhisattvas whom they worship as gods and goddesses. In addition, they adopted deities from other cultures and indigenous religions, use sacred objects and perform religious rituals and rites. Over time, I have heard many names of gods and goddesses, witnessed religious festivals and taken part in pujas, but I realise that I will remain an outsider; a visiting 'Peeping Tom', attracted by the colourful and exotic shows.

The devotedness of the woman prostrating in front of a photo showing the Dalai Lama tells me once again that I do not belong here. I slip out quietly, leaving behind a donation for the monastery.

This part of the Manang valley is arid and dominated by cliffs of yellow rock, eroded into spectacular pillars. Just behind the monastery, I see a sandstone pillar with a big hole through which the blue of the sky is visible. The wind drives shadows, cast by a few small clouds, in a wild race across the landscape, and the craggy rocks and stone pillars seem to become alive in the light of the late afternoon. I cannot help thinking of soldiers on guard, protecting the monastery.

A glance at my watch tells me that it is time to return to Manang; time to go to the cinema.

DAY 85
MANANG – THORONG LA HIGH CAMP
iPod day

Most tourists who are on the way to Thorong La stay at Ledar to avoid problems with high altitude but, being perfectly acclimatized, I decide that my destination for today will depend on my mood. I leave the hotel at six o'clock in the morning and, taking pleasure in the quiet and peaceful atmosphere of the early hour, I wander through Manang. Bluish smoke is wafting through the narrow streets and the smell of

Early morning in Manang

burning junipers mixes with the smell of spicy food locals prepare in their houses. Two women sweep the main street and cause swirls of dust and dirt to dance in the air. I quickly turn away to avoid breathing in the blend of dust and dried manure. Beside an old stupa, a man is busy loading his pack animals, and I wonder whether he also wants to cross the pass today. Locals have told me that they manage to cover the distance between Manang and Muktinath in one day, but that trekkers may need two or three.

It virtually never happens that I dig out my iPod from the depth of the rucksack to listen to music, but today I go for a full session: Genesis, Mozart, Column Sands, Vivaldi, Dire Straits and Fauré. Admittedly this collection is weird, but perfect for a day like this. Genesis for steep ascents; Mozart while wondering about the perfection of nature; Column Sands while pondering about life; Vivaldi for flat sections; Dire Straits because I deem Mark Knopfler one of the greatest guitarists of all time and Fauré to dream along during prolonged rests.

I make good progress in spite of numerous breaks at restaurants along the trail... a cup of tea and an omelette here, some coffee and a pancake there... At one thirty p.m. I reach Thorong Phedi, the starting point for the majority of tourists who want to cross the pass. There is another hotel higher up, Thorong Phedi High Camp (4,800 metres), but to reduce the dangers of high altitude sickness this hotel cannot be recommended to trekkers coming directly from Manang. Feeling perfectly fine, I tinker with the idea of walking on but, before doing so, I need food as I am starving.

Like all the villages and hotels along the Annapurna Circuit also Thorong Phedi has changed. The old hut from the '80s that resembled a cosy cow shed is gone, and a new comfortable mountain lodge now caters to the needs of tired and hungry trekkers. The warm and cosy atmosphere attracts me and promises a refreshing break, the menu, however, attracts me even more! It reads like a heavenly narration: an enormous variety of pasta dishes, various potato dishes, pizza, moussaka, Mexican dishes, all sorts of soups…

The possibility of a second helping makes me go for Dhal Bhat once again, and I do not regret it. The owner of the restaurant serves a giant mound of rice, delicious vegetable curry and spicy pickles and places a large bowl filled with a tasty lentil soup beside the plate. Even before I manage to finish the first serving, which was generous enough to feed two trekkers, he asks me if I want more! How polite he is. 'Well, yes, please. I want more, more of everything'. This answer seems to take him by surprise, but two minutes later he reappears with a tray and the same amount of food I had before. Great! That's perfect! While I finish the second helping, he watches me with a mix of curiosity and shock in his eyes, but he does not comment on my appetite.

After this refreshing break and the good food, I ramble on and reach the High Camp Hotel one hour later. Admittedly, I have to wait a bit before ordering Dhal Bhat for dinner.

DAY 86
THORONG LA HIGH CAMP – MUKTINATH
Prince Nicely and the Woman They All Talk About

From Thorong La High Camp, it is a mere 600-metre climb to the pass and there is no need to hurry. At seven o'clock in the morning, the hotel is still enveloped by thick mist, but as the trail is wide and cannot be missed, I decide to set out. It does not take long until the visibility improves and it is then that I see another tourist, probably 100 metres above me. Shall I try to catch up with the man? No. What for? Today's rhythm is, fifty minutes of walking, then ten minutes' rest with tea and cookies. This is perfect, and I do not want to change it. After my first stop, the man has disappeared into the hilly moraine.

It is not the first time I have walked over Thorong La, but it is the first time I have enjoyed good visibility and an excellent view of Thorong Ri and Khatung Kang, both 6,000-metre peaks. I spot thousands and thousands of prayer flags fluttering in the fierce wind. I visualize myself sitting in front of the small basic hut I know to be on the pass, drinking the tea the owner sells at astronomic prices. However, what does astronomic mean? Drinking a cup of cappuccino in Piazza San Marco in Venice will cost you ten Euros, and the surrounding makes you think it is worth it. Here, it is the same.

Mentally, I am prepared to invest a couple of Euros in a cup of tea, but to find the hut closed... I am disappointed. All of a sudden a miracle happens – at least if you are a woman who has been a member of the Club 40 Plus for too long. A young, handsome man appears from behind the hut, hurries towards me with widely outstretched arms, gives me a hug and kisses me on both cheeks. 'Congratulations! You are on top of Thorong La', he announces cheerfully and adds, 'You must be the woman the people in Thorong Phedi talk about'. I stare at him, baffled and speechless for a few seconds, but then I ask 'Oh, people talk about me? How's that?' He bursts out with laughter and replies. 'Down in Thorong Phedi they talk about a woman who had a late lunch there – eating soooo much', and he draws a massive mountain in the air. Being famous can have many reasons. I am famous for the amount of food I manage to eat. This is quite all right.

After the usual trekker's small talk, we go separate ways. Regarding age, Prince Nicely (that's how I call him) could be my son, and I would not be able to cope with the speed he starts running down the slope, but we agree on meeting in Muktinath to celebrate the safe and successful trip over Thorong La.

Left: Sacred springs in Muktinath **Right:** Our sadhu

In the Tibetan language, Muktinath is called *Chumig Gyatsa*, which means 'the sacred place of a hundred-odd springs'. All devotees coming to Muktinath are expected to bathe in the cold water that is believed to flow underground directly from the sacred Lake Manosarovar near Mount Kailash in western Tibet. Like many other sites for pilgrimages, Muktinath is revered by followers of both religions. Though most pilgrims are Hindus, Buddhist nuns are in charge of the entire complex. This shows, once again, the remarkable tolerance that visitors find all over Nepal.

DAY 87
MUKTINATH – KAGBENI
Followed by a Sadhu

Prince Nicely left Muktinath in the early morning, but since it will take me less than four hours to stroll down the hill to Kagbeni, I opt for a prolonged breakfast before setting out. Along the trail, a number of cosy restaurants and cafés cater for the needs of hungry and thirsty tourists and pilgrims, and it is hard to resist the temptation of stopping for tea or coffee almost everywhere. In one of the garden restaurants, I meet a young German man with a sadhu in tow.

The majority of sadhus are from India, and ordinary people refer to them as Babas. They are renunciates who have left behind any material or sexual attachment and now live in temples, caves or forests. Hindus meet sadhus with utter respect for their holiness, but, at the same time, they fear their curses. Many devotees believe the ascetic practices, performed by sadhus, help to burn off the community's karma, thus benefit society in general. This is the reason why many people support sadhus, be it with money or food.

The urban populations of India and Nepal, however, meet sadhus with a certain degree of suspicion. Especially in popular pilgrimage cities, beggars see posing as a sadhu for tourists as a source of earning money. 'Our' sadhu seems to be one of them. Yesterday evening I saw him, together with other sadhus, returning from their 'workplace' beside the entrance to the monastery. They had been colourfully dressed, and with their brightly shining begging bowls stuffed away in their bags, they had walked into one of the new hotels in the village. Today, right after dawn, they were on their way back to work, neat and clean with their bags tugged under their arms.

We are followed by our holy man, whom I suspect of not being holy at all, into every single restaurant and café. Obviously, he expects us to invite him for food and drink. Again and again, we try to explain that we neither desire his company nor need it, but he sits down beside us and smiles. Regardless if our voice reveals anger or friendliness, he smiles and I become furious with his smiling face. His indifference to our reactions shows superiority that infuriates me even more, making me realise how much 'power' he has over us. Arriving in Kagbeni, my patience is wearing thin and, when the sadhu follows me into the hotel I ask the owner for help. He guides the 'holy man' to the front door; friendly but resolute.

My hotel room faces towards Upper Mustang, and the peaks of the Himalaya, especially those of Nilgiri in the Annapurna Range, provide a striking backdrop to the town. The late afternoon light makes the landscape look like a watercolour painting, a scenery of unsurpassed beauty, but, as the sun drops slowly, the colours change until all that are left are shades of grey...

DAY 88
KAGBENI
Rest day

I wake up at five thirty a.m. when the first rays of sun find their way through the gauzy curtains and touch my face. The hotel is ideally situated in the upper part of Kagbeni and offers an unobstructed view towards Upper Mustang and the vast, dusty valley of the Kali Gandaki which flows down from the Tibetan plateau. The landscape reminds me of pictures I have seen showing the Grand Canyon. Also here, erosion and a river have formed rugged mountains and the steep walls, lining the valley, are bare of vegetation.

Historically, Kagbeni was a mighty fortress town on the southern border of the Kingdom of Lo. Situated between green fields, it bears a resemblance to an oasis in the middle of a desert. The people of this region, called Loba, are ethnically Tibetan and speak a Tibetan dialect. For centuries, their life has been determined by Tibetan Buddhism and today, the Loba community is one of the best surviving examples of traditional Tibetan culture in the world. Due to the strategic location of Kagbeni, along the trade route between India and Tibet, the Lobas experienced moderate wealth in older times, but prosperity has dwindled after the trade had come to a halt.

Kagbeni

Even in a faraway place like Kagbeni, things are changing. Khenpo Tenzin Sangpo, the abbot of Kag Chode Thupten Samphel Ling Monastery in Kagbeni, is worried about changes and writes:

'The main thread I find so far is a lack of awareness and understanding among the local people about the value and benefit of their culture. The adoption of Western consumerist culture has also posed similar threads among the youths, making them more attentive and exiting in obtaining a lifestyle with more, or only, material luxury'.

'How true, how true', many Westerners may think, but do we have the right to criticise people whose existence has been characterised and ruled by deprivations? Can we blame them for longing for a lifestyle we have confronted them with for decades? A lifestyle they have seen in Indian and Chinese adverts shown on TV? Education would help to make the local people aware of the disadvantages materialism brings, but they are denied education by their own politicians.

Yesterday, I phoned Nilam, my agent in Kathmandu, and asked about the permit for Upper Dolpo. I learned about some delays in the process and that Temba and three porters will not arrive before tomorrow or the day after tomorrow. Good, this gives me more time to explore Kagbeni and dive into a medieval world with modern cafés and restaurants.

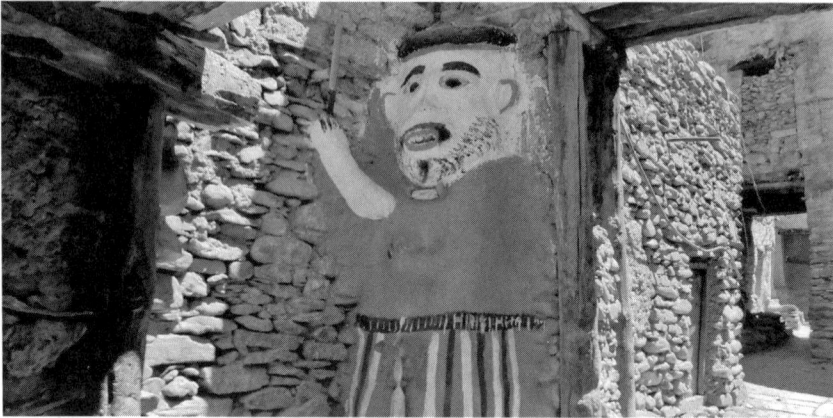

Protector statue

DAY 89
KAGBENI
Waiting for Temba

In 1987, I spent Christmas in Kagbeni and celebrated the day with a steaming hot noodle soup for breakfast. I was sitting in the dining hall with my sleeping bag wrapped around me, wearing a down jacket, a woollen hat and gloves, it was ice-cold outside and inside. I still remember the deep shock I experienced when asking for the toilet because the answer had been 'everywhere'.

Today, twenty-five years later, I stay in luxurious accommodation with an en suite bathroom where hot water gushes out of the showerhead twenty four hours a day. Whereas the people in Kathmandu experience power shortages on a regular basis and have to live without electricity for eighteen hours a day, I can plug in or recharge my electronic devices all day long, have a free-of-charge wi-fi connection that allows me to communicate with my friends at home, and in the entrance hall, could watch TV if I wished (I don't – thanks, but no thanks, I have no need for entertainment).

Temba and the porters are on the way, but monsoon season does not mean rain only but also landslides, delayed buses, jeeps that get stuck and other minor inconveniences. Their delayed arrival gives me one day more to hang out in Kagbeni. Unfortunately, July is off-season, and since hardly any tourists drop by, many restaurants and cafés are closed. The whole village seems to be deserted, and so it takes me by surprise to meet a small group of travellers in one of the narrow northern lanes, gathered around a protector statue which is amongst the main attractions in Kagbeni. Actually, it is more a relief, showing a strong man whose job it is to scare evil spirits away. His strength is symbolized by a long penis sticking out of the mud wall, and like on my previous visits, I cannot make up my mind whether it is sad or amusing to assess male strength by the length of his 'best part'?

The town is a maze of mud-walled buildings. Some of them are topped with coarse turrets and thus bear a resemblance to an architectural style found not only in other Asian countries but also in North Africa. Prayer flags fluttering in the wind and vultures circling on the updrafts, however, tell me that this is Nepal. To provide better stability in case of earthquakes, the thick walls of the houses are slightly tilted inwards and the windows are small; many still with old, wooden frames with beautifully painted carvings. A couple of houses are so close to each other that the lanes run through tunnel-like openings under the buildings. The architecture is awe-inspiring, and I am thoroughly impressed by the skills the Lobas had developed hundreds of years ago. Knowing that these buildings have resisted earthquakes, I do not worry about the stability, even noticing that there are only wooden beams supporting tons of mud and stones.

DAY 90
KAGBENI – KHARKA
Porters with blue nails

In the early morning, Temba and three porters for Dolpo have arrived in Kagbeni. One of them is Pimba, who joined me on the first part of my trip. I am pleased to see him again, remembering his cheerfulness and his infectious good mood. The two other members are new to me. When having a closer look at their hands and feet, I can hardly believe my eyes as their finger and toenails are blue. Reading this, you might think of frostbite, bruises or contusions – but no, the explanation for the colour is far less painful; blue nail varnish.

In Temba's opinion, the equality of men and women is the key to a better future, and when putting together a group of porters, he will always give jobs to his female relatives. So, for the following weeks, two of his cousins, Mingma and Jomma, join me as porters. The blue nail polish, however, is the only thing these young women have in common. In all other respects they come from two different worlds.

On her arrival, Mingma wears a traditional Bhoti dress, similar to Tibetan attire. It consists of a long, woollen, sleeveless dress called a chuba, which is worn over a long-sleeved blouse. Usually, chubas are over-sized and, when tied around the waist by a belt, the upper part of the dress becomes a loose pocket for carrying all sorts of things – even livestock and babies! Jomma, on the other hand, looks like she has come from a photo shoot for a fashion magazine: stylish 7/8 pants, a matching sleeveless top and hair accessories to match. Even her sandals were obviously chosen by taking into account the colours of her outfit. Looking down at my apparel, I feel embarrassed.

My team had been travelling for two days to join me; a strenuous journey by bus and Jeep. In my opinion, the most important things for them right now are a good rest and a good meal. The hut we want to go to is not more than three hours away from Kagbeni and, therefore, we agree to leave the village after lunch.

On our way up to the shepherd's hut, Temba points out a grassy slope in the distance, 'We call it 'soldier hill' because the Khampas (Tibetan warrior class) had their war camp here'. In 1959, when the Dalai Lama fled from Tibet, they guarded him all the way to India. Later, more than 6,000 Khampa fighters established bases in Mustang from where they crossed the border to ambush Chinese troops; supported by the CIA. Many Khampas were secretly flown to a US army base in Colorado where they were trained in guerrilla warfare. After President Richard Nixon's historic visit to Beijing in 1972, however, the US buckled to powerful pressure from China and deserted the Khampas. In the wake of this betrayal, the Dalai Lama sent a message to the Khampas, asking them to surrender and leave Mustang. Nepalese Army Gurkhas butchered those who refused to lay down their weapons.

A few things are new to me, and I am surprised to encounter traces of world history in the remoteness of Upper Mustang.

DAY 91
KHARKA – SANTA
'Illegal' in Upper Mustang

The name Mustang means 'fertile plain' and has been used for the former independent Buddhist Kingdom of Lo (meaning 'south') since the area came under the control of Nepal. But people still refer to it as Lo and talk about 'their kingdom', even though Mustang's status as a kingdom ended in 2008, following the end of its suzerain Kingdom of Nepal the same year.

Until 1974, Tibetan rebels had bases in Upper Mustang and, after that, the region remained a military zone, closed to foreigners until 1992. Even today, foreign visitors need to obtain a special permit to enter because tourism to Upper Mustang is regulated and the region 'protected' by the Nepalese government. The aim is to improve the opportunities for nature conservation, to guarantee sustainable development for the local communities, to conserve and restore the cultural heritage and to develop alternative energies. To reduce the number of tourists visiting Mustang, two restrictions have been imposed. One of them is the high price for permits (500 USD for ten days), the other a requirement to travel in a group of at least two people.

The protection project is highly controversial amongst the local public because hardly any of the many dollars paid arrives in Mustang. Why not? The problem originates in a combination of violation of law, fraud, exploitation, discrimination, sluggish bureaucracy, nepotism and corruption. As a result, local youth leaders in Mustang threatened to bar tourists, beginning on October 1st 2010. In 2011, locals began stopping tourists and denying them access to the former kingdom. Protests like these are not aimed at tourism as such, but are the reaction of Mustang's inhabitants to the failure of the Nepalese government to pass any of the money from permits to the local communities. Irrespective of the form of government, the country and its people remain in the stranglehold of policies they have no chance to change. Hardly any of the politicians genuinely cares for the needs of the inhabitants. They prioritize their own interests and there seems to be no end. Nepal is a country living on hopes.

Since we left Kagbeni, we are travelling 'illegally' in Upper Mustang as we lack a permit allowing us to enter this region – a rather worrying situation. I remember an unpleasant incident from a visit to Kagbeni in 2000 when a furious police officer stopped me, pointing his gun at me. In his opinion, I had come too close to the border when taking pictures of the Upper Mustang Valley.

There is talk about a special rule for trekkers crossing the western part of Upper Mustang when heading for Dolpo, but despite having sent several inquiries to the Immigration Office where permits are issued, I never received a reply. Thus, we can only hope that the unofficial rule exists. The fine would be exorbitant and result in bankruptcy or, at least, force me to stop my journey.

After nine hours following the steep trail up the mountain, we arrive in Santa without encountering any problems in the form of check posts or furious police officers. The village, built in an elevated and sunny place, shows features typical of Tibetan architecture. The mud brick houses stand close together as though to protect each other from the extreme cold in winter, and a few stone-walled fields hide between the buildings, thus preserving the fertile soil, which the frequently occurring fierce winds would otherwise blow away. Open courtyards serve as stables for the families' ponies and as a place to dry vegetables and fruits. The flat roofs allow the easy removal of snow in winter and help to conserve the heat as little fuel is available. Soaring piles of firewood lining the roofs are a common sight in villages along the Annapurna Circuit. In Santa, many roofs are almost empty; a bad sign. In an almost treeless area like Upper Mustang, a well-stocked wood store means that the owner of the house is wealthy and, the older the wood is, the richer he is. The amount and the age of fire wood tell others that there is no need to touch the last logs. Sometimes, there are still logs that were cut by fathers or even grandfathers. Santa, however, is poor and empty.

DAY 92
SANTA – KHARKA
Life in Mustang

The trail from Kagbeni into Upper Dolpo crosses an area rarely visited by tourists, and so income from tourism is negligible. The local people wrest a living from an inhospitable landscape by combining agriculture, animal husbandry and trade, while migration between permanent villages and pastures at higher altitudes characterises this agro-pastoral livelihood. The low standard of living has declined even more since the traditional salt trade with Tibet stopped after the Chinese occupation. The closure of the border was also an enormous blow for the local shepherds who no longer were allowed to drive their yaks, sheep and goats to the fertile summer pastures in Tibet.

We follow the deep valley the Kyalunpa Khola has carved into the landscape. On the other side, the houses of Ghok are scattered on a high plateau. Due to the general aridity of Upper Mustang, agriculture depends on irrigation systems taking the precious water to the fields. The local farmers grow mostly potatoes, buckwheat and barley and, in the more favourable climatic conditions of sheltered spots, they also grow some fruit and vegetables, but the yield is barely enough for families to survive. Even when there is a little left to sell, the farmers face a strong

1 The main lake at Gosainkund, a holy site for both Buddhists and Hindus.   2 A village in Helambu.
3 The internet café in Syabru Besi.

1 The outdoor school in Gatlang.   2 A monk from the Ribum monastery in Lho.
3 Inside the prayer room of Braga Gompa.   4 The view from Thorong La. This wasn't my first time over the pass, but it was the first time the visibility has been good enough to enjoy the view from the top!
5 The 'modern' Prayer Wheel in Manang, complete with Nescafé tin. **Photo** Ulrich Schroeder.
6 The children of the family we stayed with in Tipling. In typically Nepalese fashion, the children, parents and grandparents all live in the same house.   7 Bhudi Gandaki valley – the most impressive landscape on the Makalu Circuit.

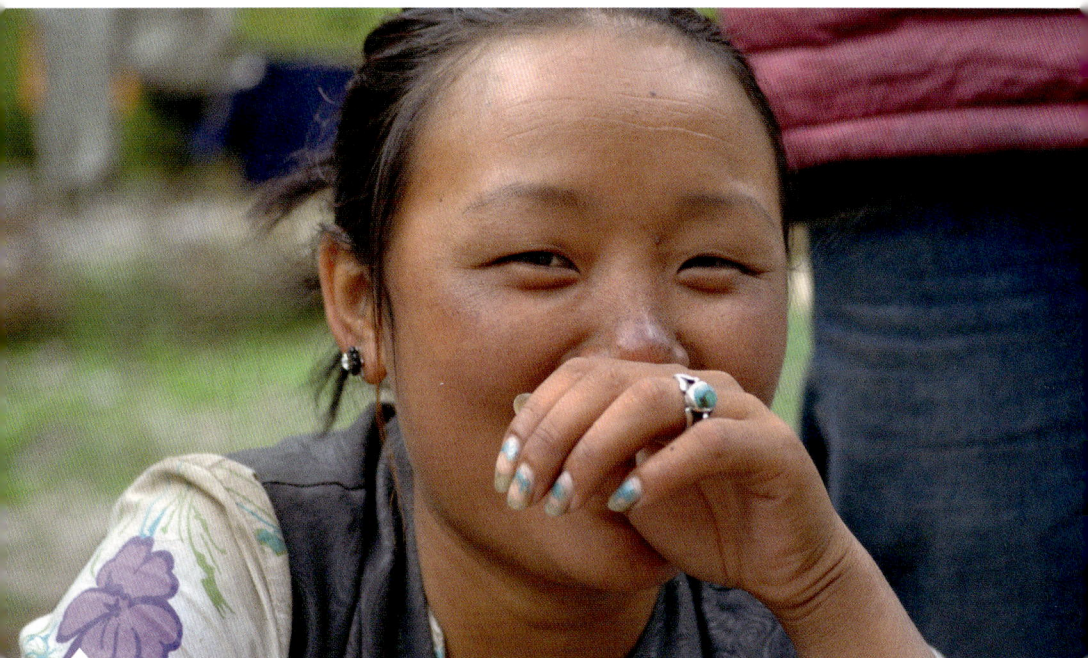

1 Jomma, Mingma, Temba and Pimba on Jungbenley La.   2 Pupils from the Crystal Mountain School in the Upper Dolpo.
3 Jomma's blue fingernails. Looking like she had come from a photo shoot for a fashion magazine, Jomma joined me as
   a porter for a week.   4 The mountains above the Kali Gandaki.

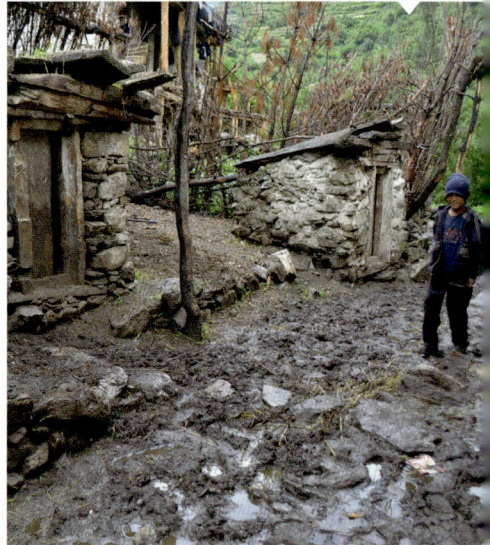

1 Shepherds' tents in the green pastures of the Thasan Khola valley.   2 The greenhouse project in Dho Tarap. Launched by the French, the project aims to combat the harsh environment and allow vegetables and fruit to be grown, thus combating malnutrition in the region.   3 Village street in Bam in monsoon season.   4 Pilgrim on the way to Shey for the annual Buddhist Festival.
5 The spectacular trail along the Phoksumdo Lake.

1 A chorten in Bhijer.

competitor; China. Following the completion of the road between Nepal and Tibet, cheap imports have begun to flood the region, including basics such as rice and flour.

When climbing down to the bottom of the valley we are baffled to see that the old, rickety bridge the guidebook describes has been replaced by a new suspension bridge. From here, the two-day climb to Jungbenley La begins, which is said to be the official border between Mustang and Dolpo, but who knows that for sure? The more people you ask, the more different answers you get. And does it even really matter? The local population does not care for, apparently, arbitrary boundaries.

A small flat area beside a creek makes a potential campsite, and we decide to stay here for the night. As soon as the tents are set up, we watch a young couple with a toddler arrive and settle down fifteen metres away from us. They unpack their few belongings, place the cooking gear on the ground and coax a fire into life to make tea. Soon, the developing smoke from the wet wood bothers the child, who begins to cough and cry until the mother soothes it by offering her breast. The poverty of the family is obvious. There is no pack animal for transporting their luggage, their clothes are worn out and insufficient in poor weather conditions, the cooking pots are old and battered and, as far as I can see, they do not even have a tarpaulin to protect them against wind, rain and snow.

Empty roofs in Santa

Yes, even in the middle of summer, snowfall is likely to occur at higher altitudes.

Shortly afterwards, the man comes over to our tents and explains that he and his family have been without food since the day before. Temba doles out rice and lentils so the family can prepare Dhal Bhat and, it goes without saying, that he offers them a dry place to sleep in the kitchen tent when it starts raining after dinner.

NINE

UPPER DOLPO/MUGU

Monastery in Dho Tarap,Dolpo

DAY 93
KHARKA – NALUNGSUMDA KHARKA
Finally, we are in Upper Dolpo

The map shows that we have three 5,000-metre passes to cross today, and I fear the worst. Surprisingly enough, none of us has any problems with the 900-metre climb up to Jungbenley La, but gazing at the course of the trail ahead, an unpleasant reality unfolds.

The path runs down to the wide, flat valley floor of the Lhanimar Khola and, from there, it winds up again in countless, steep switchbacks. In the mountains, I often fantasize about being a bird that flies from peak to peak or from pass to pass, unhampered by gravity. The desire to fight gravity tortures my mind for a while, but there is no way out; I will have to walk down and up again.

Beside the Lhanimar Khola, we encounter a small group of traders, travelling to Jomoson to purchase goods that they need at home. Their healthy-looking pack animals, who are decorated with red woollen strings and have tassels plaited into the hair, use the break to graze and drink. The traders invite us for a cup of tea and, after the usual small talk, we decide to spend our lunch break with them. My thoughts, however, wander off to the zigzagging trail starting only a few metres away from the lunch place and I worry about the hardship the long, strenuous ascent will entail. As a result, I am not able to enjoy the delicious meal.

'Never give up', are the Dalai Lama's words, and whenever life is hard, these words become my personal mantra which I repeat and repeat and repeat – as I do today. To my surprise, the ascent is not as strenuous as I had feared and, on reaching Jungben La, I am rewarded a breathtaking view to Hidden Valley and Dhaulagiri. It frequently pays to shed some 'blood, sweat and tears', but it does not pay to worry the way I did, allowing thoughts and fears to dominate life and 'steal' the joy of a good meal.

Many locals consider the pass the border between Upper Mustang and Upper Dolpo. The latter became known to the West after the Tibetologists, Giuseppe Tucci and David Snellgrove published accounts of their journeys through Dolpo in the mid-1950s. Two decades later, the zoologist, George Schaller, obtained special permission to enter Dolpo to study the Himalayan blue sheep, accompanied by Peter Matthiessen who was searching for the legendary Snow Leopard. A long time ago, I had read their books, and my old dream of visiting this region has come true today.

Outside visitors to Dolpo, primarily to Upper Dolpo, have been few in number due to complicated logistics, topography, or government restrictions. In an attempt to preserve the unique culture and ecosystems found here, the number of visitors allowed in remains strictly limited. This area, located on the Tibetan Plateau in the great Himalayan rain shadow, is considered one of the harshest and roughest places in the world, geographically cut off by a barrier of high mountains, deep gorges and dense forests.

Dolpo is often referred to as the legendary Ba-Yul or 'Hidden Land', and I wonder if Dolpo is the mysterious Shangri-La which the British author James Hilton described in his novel Lost Horizon. Of course Shangri-La is pure fiction, but it has become a synonym for any earthly paradise that is isolated (and hidden) from the rest of the world, imaginary Himalayan utopias in particular. Buddhists believe that Padmasambhavato created these idyllic and sacred places for devotees to seek refuge during times of conflict. Even the Nazis were excited about the secret land where people are always happy and virtually immortal. In the hope of finding an ancient master race, they sent an expedition to Tibet in 1938-9.

At first glance, Dolpo seems to be this 'Hidden Land' as there are high mountains surrounding remote valleys with lush, green pastures. Since the inhabitants have never been subject to Chinese influence, their traditional Tibetan culture is thriving and makes Upper Dolpo a paradise for Tibetologists. I assume that the romantic tourist will see it the same way, but what about the people of Dolpo, the Dolpopa? Do they share this romantic view? I doubt it. Even today, the number of people who can write and read is negligible, eighty per cent of the population are illiterate. Almost ninety per cent of the Dolpopa live below the poverty line and face grinding living conditions that would probably break anybody in the West. Most villages have neither medical facilities nor safe water projects, a postal service and electricity are non-existent. As far as I know, there is not a single doctor in Upper Dolpo.

Still, we need to cross one more pass, the Niwas La, but it turns out that it only can be defined as a pass when coming from the opposite direction. For us, wandering downhill to Nalungsumda Kharka, it is no more than a small, flat area.

DAY 94
NALUNGSUMDA KHARKA – CHHARKA BHOT
Easy walk to Chharka Bhot

The walk from Nalungsumda Kharka to Chharka Bhot feels like a Sunday promenade as we follow the course of the Thasan Khola. While strolling down the valley, we soon forget the toils of the previous days. The first shepherds have arrived at the green pastures we pass and, now and then, we share news over a cup of butter tea at their tents. They will spend the summer months here in the maze of valleys, wandering around with their yaks and horses to find the best grass. It is an idyllic area with numerous brooks, creeks and streams coming down the from side valleys. The trickling, gurgling and babbling sounds are music in my ears; elation grips me and I want to dance down the valley.

The village of Chharka Bhot, situated at an altitude of approximately 4,300 metres at the confluence of the Tsang Chen and Tsang Chung (Big River and the Small River respectively), consists of two parts. The new village houses a few hotels, campsites and shops, and colourful billboards in English try to attract the attention of the hundred or so foreigners who come through the settlement annually. Even though there is no signal here, these billboards give mobile phone numbers for their wares!

The old part of the village resembles a compact, medieval fortress where the houses stand even closer together than in other villages in Dolpo. Apparently, the narrow valley limits the amount of arable land, so the locals have chosen to limit sprawl. The houses' yak pens form a kind of wall around the village and there is one entry only to admit visitors. It comes as a surprise to find some three-storey buildings in Chharka Bhot because

Wood transportation

high-rise buildings had been reserved exclusively for the local nobility. While exploring the medieval lanes, I wonder about life in this remote village and whether it has changed significantly over times; I doubt it.

In this region, wood is rarely used for construction because the landscape is barren and trees, let alone forests, are few and far between. Caravans of men, horses, donkeys and yaks transport the material from distant places to this region, and since the condition of the terrain is often extreme, long loads, such as wooden pillars, have to be carried by hand. In some places the paths are so small that it is even difficult for men to traverse. The weather may also complicate the success of the caravan, not to mention its survival. These journeys are, doubtlessly, demanding and often dangerous.

The Nepalese people use the word *bhot* to describe a frontier village of Tibetan culture at the fringe of what they consider civilisation. In all likelihood, we would need five days to reach the nearest Nepalese town that is connected to the road network. The locals, however, talk about as little as two days, and this fact probably explains why the inhabitants feel closer to Nepal than to the even more inaccessible depths of Dolpo. They do not even bother to attach the suffix *bhot* to names of other Dolpo villages considering them beyond the border.

DAY 95
CHHARKA BHOT – CHAP CHU
'Real good trekking'

Early in the morning, we set out to see a Bon Gompa in the vicinity of the village, but unfortunately nobody is there to unlock the door and so

River crossing

we decide to walk on. There are two routes connecting Chharka Bhot with Dho Tarap, and we opt for the adventurous southern route. The trail runs down a sandy slope to the confluence with the Chuchen Khola and follows the Chharka Tulsi Khola, which has to be crossed several times. The fact that there are no bridges does not cause problems as such. Like us, the locals take the trail via Chap Chu because the northern route is not suitable for the pack animals they normally take along to facilitate transport. Thus, we just have to follow their footprints when searching for the best places to wade through the river. Unfortunately, we did not take into account that the Dolpopa rarely wade through rivers but ride horses or yaks. They only get wet feet whereas we get soaked up to our thighs.

A couple of times, we discover that the footprints we are following belong to grazing animals that swam through the river. We are not keen on taking a swim, but finding fords is not as easy as envisioned, and soon we know why this route is described as hard or even dangerous in, or after, rain. Several times, we are forced to retreat because the water is too deep or the current too strong – or both. Smooth stones and rocks are an additional risk factor, and I am glad to have my walking poles with me. More than once, Temba has to ford the river a few times, either to give my poles to Jomma and Mingma, or to offer them a helping hand. We consider ourselves lucky because the sun is shining, and the water is not ice-cold. But even so, water draws heat away from the body faster than air does and after a few hours we can feel our body temperature beginning to drop. In bad weather, this 'river trip' could easily result in hypothermia. Temba is beaming with joy all day long and cheerfully exclaims several times 'This section is good trekking, really very good trekking!' I totally agree with him.

Despite the excitement, which we all thoroughly enjoyed, we are happy when we finally leave the valley and climb up to Chap Chu, where we will stay for the night. Chap Chu is not a village but a vast grazing ground beside a tiny lake; a hugely popular place for nomads and shepherds who come here with their animals during the summer. They have established a tent village, and we are surprised to even find a tent shop where we can buy fresh butter and sak*.

DAY 96
CHAP CHU – DHO TARAP
Modern times in Dho Tarap

Today, we have to cross the Chan La (5,378 metres), but since all of us are fully acclimatised, the moderate climb does not pose any difficulties. The trail repeatedly crosses back and forth before it takes us to a wide basin from where we, finally, continue up a gentle gradient to the pass. A fierce, cold wind greets us, and so I do not spend much time looking at the magnitude of the peaks around us. Endless switchbacks take us down into the valley of the Tarpi Khola and several hours later the first houses of Maran appear in front of us. At last! I am almost dying of thirst after our long walk in the dry air of high altitude. Temba finds a family who has time to make tea and Tibetan breads for us. Contrary to chapattis, which are cooked on a griddle, Tibetan breads are fried in oil. They are delicious when freshly prepared, but their texture quickly begins to bear resemblance to old leather as soon as they get cold. Dho Tarap, our destination for today, is no more than a thirty-minute walk from Maran, and so there is plenty of time to enjoy a prolonged break, with several crispy Tibetan breads for each of us.

The sun has already disappeared behind the craggy, rocky ridge when we set out again. While the last rays of the sun bathe the grey stone houses and lush green fields on the other side of the valley in a warm, glowing light, we walk in the shadow of the mountain. Strong thermal winds drive the afternoon clouds at high speed across the sky, creating dramatic interplays between light and shadow on the barren slopes.

Dho Tarap houses a campsite beside the river, a well-stocked shop and a police station that is fully staffed during the Yartsa Gunbu season. We have just settled down when three police officers pay us a visit. I assume,

---

* Sak is a vegetable similar to spinach which is served with every Dhal Bhat.

they come over to scrutinize my papers and permits, but they do not even ask for the second tourist who should be with us, so I am allowed into Upper Dolpo as a group. They just come over for a chat and a cup of tea because they are apparently bored (or even annoyed) about having been posted to the middle of nowhere. Judging from their appearance, they are not from this region. The local people resemble the Khampas from Tibet. Men and women alike wear their hair long, but only the men tie their hair back in a ponytail and decorate it with red, woollen tassels. The majority of women wear necklaces made of corals, turquoises and musk deer teeth. The latter hints at poaching, but the police do not seem interested.

DAY 97
DHO TARAP
Rest day

Large areas in Upper Dolpo are still in the firm grip of ancient times and arriving in Dho Tarap is like travelling back in time to a medieval village. The observant traveller, however, will soon spot clear signs of the twenty-first century. One of the villagers, Jyampa Lama, owns a flat screen TV powered by solar power. And he is not the only one to invite friends round for a TV session to watch football or wrestling – even the local lama, Amchi Namgayal Rinpoche, has TV in his monastery. He, however, does not only see the advantages of being connected to the outside world, 'Now, the people have some money, and they prefer buying consumer goods instead of making them themselves. In older times the people used to make their own shoes, but nowadays, craftsmanship is no longer of any interest. Tourists wear modern clothes and shoes, and when the locals have money, they long for the same things. As soon as there is money a need for something will turn up'. Like Khenpo Tenzin Sangpo, the abbot from the monastery in Kagbeni, he thinks that it is of uttermost importance to keep one's own culture and to adopt only the positive things from the West, not the negative ones.

My clothes have reached (and probably exceeded) saturation point for dirt and smell. Washing tops the to-do list for today. The cold water, gushing out from the local water tab, is not sufficient to remove the dirt completely, but an overdose of washing powder helps to offset the unpleasant smell to some extent.

An old Bon monastery is situated above the village, and I trudge up the hill to visit the gompa. To my disappointment, neither the Amchi Lama nor the old woman who is said to hold the Lama's key in his absence are present and so I decide to wander along the green valley to the school of Dho Tarap; the Crystal Mountain School. It is regarded as a model school in many ways, and I hope to meet people there who are interested in hearing and learning about autism and my project.

In 1994, just two years after Dolpo had been opened for tourists, the French organization Action Dolpo founded Crystal Mountain School. In its founding year alone, thirty pupils attended the boarding school, which was the first school ever in this region. Today, almost twenty years later, 130 pupils undergo an eight-year education, and those who think about higher education can attend a boarding school in Kathmandu, financially supported by the French. Action Dolpo has ambitious targets for their project, but since a lot has been achieved already, I am very optimistic about the future.

From Grade 1, the children learn Tibetan, Nepali and English, and traditional Tibetan culture plays a vital role in the pupils' life at school. The mother tongue of the Dolpopa is Tibetan, but the teachers, sent by the government, neither speak nor write this language. This causes problems and conflicts. Losing one's native language means losing parts of one's culture. Action Dolpo's ambitious aim is to preserve the Tibetan language and, therefore, they provide the necessary financial means to pay the wages for Tibetan teachers.

It is not only the school that makes a difference to the people's existence here. Tibetan medicine has been practiced throughout the Himalaya and the Tibetan Plateau for centuries. Presently, this healing tradition is undergoing intense changes. Tourism, emigration, development programmes, government health policies and market forces have deeply affected the lives of local doctors (amchi) and community health. The project wants to keep traditional Tibetan medicine and thus avoid losing knowledge accumulated over hundreds or even thousands of years. Today, young people interested in traditional medicine can study the subject at a specific medical school supported by Action Dolpo. However, since the French organisation focuses on combining the past with the future, this training is open for everybody – previously only the son of an amchi had the right to become a traditional doctor.

The headmaster of the Crystal Mountain School welcomes me and shows serious interest in autism. More teachers join our conversation and soon a discussion is in full swing. By the look of it, there are two children attending the school who show some signs typical for autism. When I tell them about the courses for teachers run by *Autism Care Nepal* in

Kathmandu, they decide to send one person working at the health post to the capital.

Before returning to the village, I am allowed to take over one of the classes for twenty minutes. I really appreciate this offer, and take the children out to the school yard, where I hold a small lesson with music and movement activities. Even after having been at higher altitudes for many weeks now, the combination of singing, jumping and running at 4,000 metres leaves me gasping for oxygen, my heart pounding. However, we have a lot of fun together.

DAY 98
DHO TARAP – DANIGAR
Greenhouses for Dolpo

Yesterday, when visiting the school and talking to the teachers, I learnt about Dolpo, the Dolpopa, Dho Tarap and Action Dolpo. Of course, I also wanted to know more about schools in other regions of Dolpo. The headmaster's facial expression did not conceal his concern and worries when he explained that in the hamlet of Shimen, a two-day walk from here, the majority of the parents were not able to pay the school fees of thirty-five USD for six months. A specific regulation gives them the option of paying with fifty kilograms of wood or yak dung. 'We who live in Dho Tarap are quite aware of the fact that we have been very, very lucky', he adds.

One of the positive things Dho Tarap has taken from the West is a greenhouse project launched by the French. Today, more than half of the households in the valley of the River Tarap, which we follow on our way to Danigar, have greenhouses. We can see the plastic-roofed buildings everywhere.

The Dolpopa live on a diet that lacks variety. The majority of their calorific intake comes from barley products, which is a good food staple. In addition to barley, the locals grow potatoes, buckwheat and radish, and they get milk, butter and meat from their animals. Yet, the harsh climate makes it impossible to grow vegetables and fruit to supplement their diet with vitamins. This leads to malnutrition, resulting in reduced body size, lower life expectancy and increased vulnerability to disease. Since disease reduces nutrient absorption but increases the energy expenditure of the immune system, the nutritional status will get worse. Thanks to this project, this vicious circle will hopefully be brought to a halt.

Action Dolpo provided the building material for the greenhouses and, this year, the local people are starting their first trial run with spinach, cabbage and broccoli. This means that a diet that has existed for centuries is about to be changed. Can an innovative project like this become a success? In my opinion it will, because by now many locals are directly involved in Action Dolpo and its partner organisations. It is no longer the 'strangers and foreigners' who come with new ideas and tell the local people what to do and how to do it, but the Dolpopa themselves. Action Dolpo focuses on long-term sustainability, not quick fixes. The majority of young people who once received their primary education at the Crystal Mountain School returned to their villages knowing that there was a job waiting for them. It is they who will change the way of living and, being locals themselves, they are welcome, respected and accepted by the community.

The greenhouses not only play a crucial role as additional arable land, but also as solar houses in the cold and long winter season. The traditional houses are without heating because fuel is scarce, but since many of the greenhouses are integrated into the main building, they can be used for warming up the house. It is a double win situation for the Dolpopa.

We continue our journey to the Numala La (5,300 metres), which we cross in the early afternoon. Everywhere, we meet nomads on their way to the summer pastures, families walking slowly up and down the hillsides in search for Yartsa Gunbu and caravans transporting heavy loads to the far away settlements and hamlets of Dolpo. We are tired when we arrive at the camp at Danigar just before dusk. Apart from our group, many Dolpopa came to this place for the night, and several of them have already pitched their tents. The smell of burning wood is lingering, singing and laughter fill the air and the whistling sounds of pressure cookers remind me of my hunger. Like us, they will continue over the Bangala La tomorrow.

DAY 99
DANIGAR – RINGMO
The legendary Phoksumdo Lake

The early morning rays of the sun seem to have a reluctance about them as they creep, in slow motion, over the ridge, but in the end the snow on Norbung Kang glistens and gleams above the head of the valley. Though the mountain is lower than the highest passes we had crossed on the way

Camp at Danigar

to Solu-Khumbu, it dominates the landscape so clearly that it can easily be mistaken for a 7,000-metre peak.

When coming from the barren landscape and the dominating sandy grey and brown colours of the region around Dho Tarap, the green of the pines and junipers that line the route to Ringmo pleases the eyes. How lucky we are in Northern Europe to have so much green around us. How little we acknowledge this privilege.

It is already getting dark when I hear a roaring sound in the distance. Then it appears – the massive Suligad Falls, at 900 feet they are the biggest waterfall in Nepal. A couple of minutes later, I reach the highest point of the ridge and catch my first glimpse of Phoksumdo Lake. People say that the colour of the lake stands in stark contrast to the rest of the surroundings but, right now, the land and the water form one dark grey entity; a nondescript, stale murkiness. However, even this greyness leads me to believe in legends; or is it in spite of it?

'A long time ago, there was a village in the place of the Phoksumdo Lake. One day, a vindictive female demon, fleeing from the Saint Padmasambhava, happened to pass by the village. Hoping that the locals would not give her away to Padmasambhava, the demon gave the village people a shiny turquoise. When Padmasambhava came into the village, he sensed that the people were hiding the demon and turned the turquoise into a chunk of dung. Having lost the stone, the village folk saw no reason to protect the demon any longer and they revealed her whereabouts to the Saint. Feeling betrayed by the villagers, the demon was angry and caused a massive flood that drowned the entire village'.

Legend has it that the remains of the village can still be seen today below the lake's surface; the brilliant aquamarine, turquoise colour standing as a reminder of this mythical incident. A more scientific approach to explain the phenomenon of the deep aquamarine hue of the water is made by taking into account the enormous depth of the lake – about 650 metres.

DAY 100
RINGMO – PHOKSUNDO KHOLA CAMP
A special day?

One hundred days on the go. This is certainly worth a celebration, but though the landscape is breathtaking, the village is not of particular interest to me. We decide to walk on.

In the early hours of the new morning, the scenery with its surrealistic components is almost scary. Lake Phoksumdo, lying in front of me, still resembles a large pond full of gloomy greyness, just as it did yesterday. But as the sun creeps over the ridge, bathing the land in a bright and golden light, the colour of the lake gradually changes from blue-green to turquoise and aquamarine creating a stark contrast with the shades of green of the fields and meadows. Just beside the water the almost vertical walls of the snow-covered peaks of Kanjeralwa and Sonam Kang rise into the sky and form a natural barrier. It is as if they want to protect a hidden treasure.

I cannot help thinking of the emerald-green and picturesque Königssee near Berchtesgaden in the Bavarian Alps which is, without doubt, one of the main tourist attractions in southern Germany. Whereas Königssee is famous for a pilgrimage church located on a peninsula, Phoksumdo Lake is known for a Bon monastery rising beside the water; the Pal Sentan Thasung Chholing Gompa. Even the Japanese tourists Bavarians usually associate with the Königssee have arrived here. I feel at home. A truly memorable day indeed.

The Pal Sentan Thasung Chholing Gompa is said to be more than sixty generations old. Many people consider it one of the main centres for Bon religion in Dolpo and, probably, the last concentrated quarters of Bon culture in Nepal. There are many speculations around the Bon religion, which preceded Tibetan Buddhism, because its history is difficult to ascertain. The earliest surviving documents referring to the religion originate in the ninth and tenth centuries, when Buddhists had already begun to suppress indigenous beliefs and practices. The result

Morning at Phoksumdo Lake

was a long historical rivalry between the Bon tradition and Buddhism in Tibet. In the beginning, the Bon religion showed similarities with animalistic, shamanistic religious forms but, like all religions, it has changed over time, and the fourteenth Dalai Lama has recognized the Bon religion as the fifth principal spiritual school of Tibet along with other schools of Buddhism.

There are many distinct practices within Bon religion, but they are barely noticeable without being a scholar or, at least, a practising Buddhist. However, seeing a local walking around a monument in an anti-clockwise direction one can be sure about his or her denomination. Buddhists circumambulate them clockwise. The habit of sticking out the tongue as a sign of respect when greeting another person originates from a time when followers of Bon religion were considered 'enemies' endangering Tibetan Buddhism. They were forced to stick out their tongues so one could find out if the person recited mantras constantly, which was believed to cause the tongue to become black or brown. This allowed the 'real' Buddhists to pick out the Bon-pos, who were suspected of reciting magic-mantras.

The path along the lake is one of the most spectacular ones in Dolpo, but it no longer bears resemblance to the treacherous path described by Matthiessen and Schaller, or that shown in the film *Himalaya*. In the beginning, it climbs up about 300 metres, and we traverse the cliffs on the western shore along a precipitous trail. Though the track is well-maintained, it feels exposed in some places. From this vantage point, the view of the lake is unsurpassed and the changing colours of the water add to the beauty of the magnificent view. A day like this one adds nothing but joy to a trekker's life. The trail to Shey Gompa, however, follows the Phoksumdo Khola as it runs out of the mountains and we have to leave this paradise after our stop for lunch.

This morning in Ringmo, we met a Japanese trekking group of four who, like us, are on their way to the Shey monastery. Their guide is one of Temba's friends, and it goes without saying that we will camp together and celebrate tonight. This offers a unique opportunity to experience luxury adventure trekking since Japanese people are known as perfectionists, with every detail of their trips being just so. These four tourists have a personal guide each to assist them in case of problems or dangers, and the amount of food is sufficient to satisfy the appetite of a whole platoon. Several horses carry bulky bags containing rice, noodles, flour, lentils, jars with jam and honey, bread, cakes, sweets, potato crisps, cabbage, carrots and more. There is even one porter just for carrying the eggs. All I can do is to wonder. This kind of luxury trekking cannot be compared with my 'travel light' method, and for a few minutes, I must admit that I envy them, if I am honest.

DAY 101
PHOKSUMDO KHOLA C. – PASS CAMP
The way to Shey?

We leave the camp with the Japanese and follow the track through the narrow valley of the Phoksumdo Khola. Locals had told us about a shorter route to Shey running along the bottom of a spectacular gorge and, for a while, we tinker with the idea of taking this 'shortcut'. Upon reaching the confluence, Temba and Pimba try to find a suitable trail between the vertical cliffs and the meandering river, but after half an hour they give up. We decide to follow the Japanese who, due to their horses, took the main trail.

The number of locals we encounter on the way amazes me, but then I remember the annual religious festival in July the Buddhists celebrate in Shey; everybody seems to go there for it. However, it is not the festival alone that makes the people set out on a certainly arduous journey over high passes to the monastery. The coming together is an excellent opportunity to meet other people, exchange news, cut deals and celebrate with friends and relatives one has not seen for quite a while.

In addition, the year 2012 is of particular importance for Buddhists.

According to Chinese astrology, it is the Year of the Dragon, and there will be a festival in August that is held only once every twelve years. Hundreds of believers will circumnavigate Crystal Mountain, their Nepalese Mount Kailash, to achieve spiritual merits. The Japanese trekkers are Buddhists, too, and their destinations are Shey and Crystal Mountain. After lunch, we set out without them. Their journey started in 'lowland Nepal' only a few days ago and, in order to avoid high altitude sickness, they must set up their camp at the lunch place.

In the afternoon, light rain starts and accompanies us for the rest of the day. The sky is grey, and I doubt we will catch a glimpse of Crystal Mountain when crossing Nangdalo La tomorrow. What a pity! The name Crystal Mountain derives from the fact that the peak is embedded with crystal deposits that glint in the sun. Together with Shey Gompa, the monastery at its base, the mountain rates amongst the great treasures of Dolpo.

We are tired of walking in the rain and search for a flat, grassy spot to pitch tents, but it turns out to be quite difficult to find a suitable campsite. Just before getting lost in the thick layer of white clouds gathering in from all sides, we get to a tiny basin that is perfect for this purpose.

DAY 102
PASS CAMP – SHEY GOMPA
The right way?

All night long, heavy rain fell, and in the morning everything feels wet and damp in my tent. Although it is the same in the kitchen tent, all members of the group are in a good mood when we leave the camp and set out for the pass. Their singing and laughing is infectious and soon I forget about the damp clothes I had to pack into the wet rucksack.

Yesterday, I had taken out the GHT guidebook a couple of times to read the route details again and again, but I could not help feeling that something was 'wrong', even though I was unable to put the finger on it. Was this the way over the Nangdalo La to Shey? Temba was dead sure. His friend who guides the Japanese group is remarkably familiar with this area, and before we left the Japanese behind, he had given Temba his GPS watch with the stored tracklog which would guide us through the remoteness of this region. Yet, I doubt and, once again, the guidebook and the map get dragged out from the rucksack. 'I am not sure, but a few things do not fit into the picture. We have not seen the waterfall mentioned in the book, and we have not come to a wide basin', I explain, but Temba relies on his friend and his friend's watch. Well…

On reaching the highest point, we compare the given altitude from the map with the GPS. There is a variation of about fifty metres, not enough to worry about. Maybe, my female intuition just played tricks on me? It does not take long before the valley unfolds below us, but we are taken by surprise to see that the valley runs down to our left-hand side. According to the map and the route details, it should be on the right side. Strange, strange. Once again, we consult the map, check the altitude on the watch, look at the terrain and, finally, we come to the conclusion, correct valley but wrong pass. No reason to worry. Let's turn left then.

Later we learn from local people that there are several points were it is possible to walk over the mountain range. The route Temba's friend had saved to the GPS watch is a trail the Dolpopa prefer, however, it is one not shown on the map.

DAY 103
SHEY GOMPA – BHIJER
Sick in Dolpo

The trekking days in Dolpo are long and physically demanding due to the high passes and deep valleys separating the settlements from one another. We decided yesterday to spend our planned rest day in Bhijer instead of Shey Gompa because, in our opinion, Bhijer is the better place for relaxation and for stocking up provisions before setting out for the last 5,000-metre mountain passes on the way to Mugu.

The climb to the ridge above the monastery is steep, but provides an excellent view towards the plain where hundreds of people had

Left: Devotees circumambulating the monastery   **Right:** Shey Gompa

transformed the tranquil and peaceful pasture into a bustling tent town. The bluish smoke of burning junipers lingers in the clear air of the early morning, men rush to and fro, unloading or reloading the pack animals, bartering and haggling while women wash blankets, feed their babies and prepare food. Children's laughter reaches our ears, and we watch them running around between the tents. In the vague hope of catching a glimpse of Crystal Mountain, my eyes search the horizon but once again, dense clouds obscure the peaks and my dream remains a dream. At least some blue patches are beginning to peek through the cloud, promising a pleasant walk in dry weather.

After a prolonged break for lunch beside a creek, the weather deteriorates suddenly. It is cold, and heavy rainfall sets in. Without our rain clothes, we would be soaking wet within a few minutes, but luckily, we are well-equipped. When having a short stop in the lee of a huge cliff, a local family catches up with us; father, mother, two boys, aged five and seven, and a twelve-year old girl. The parents lead the two horses carrying their sons who seem to be tired and exhausted. The family stops beside us and tells of the odyssey they have just made. Three days ago, they had left their small farm north of Saldang because their youngest son suffered from severely infected tonsils, high fever and loss of appetite. The health post in Saldang was closed on their arrival, and the lama, though trained in herbal medicine, could only give them something to reduce the pain. Luckily, he noticed and understood the seriousness of the situation and advised the parents to move on to Bhijer where there is a well-equipped health post.

To Westerners, this may sound almost unbelievable and unreal, but in the remoteness of Dolpo, it is everyday life. The cold and damp weather depletes not only the family members of their energy but us as well, and as a result we have a couple of short breaks and thus we meet each other several times in the afternoon.

It is our group that finally arrives first in Bhijer. Our tents are already pitched, tea is ready and rice is cooking in the pressure cooker when the family arrives at the village, totally exhausted. Since the two boys were about to fall down from the horses, the parents had decided to carry them in their arms. Now, they are completely done. We offer them tea, hot juice and biscuits before they travel on to the health post.

DAY 104
BHIJER
Rest day

My rest day starts with a late breakfast as I linger in the warmth of the sun. Temba and Pimba leave to explore the village, trying to get hold of some goodies. Unfortunately, there are no shops in Bhijer and the only person who has got a few things to sell is the local lama. At least it is possible to buy biscuits, and get some spinach from the lama's garden. After returning from their stroll, they tell me that all of us are invited for a cup of tea in the afternoon.

The monastery was built many centuries ago and is one of the oldest in the Himalaya. Due to extensive renovation, paid for by a rich French couple, it is in excellent condition. Just beside it there is another new building which I initially mistake for part of the monastery. It has the beautiful and typical Tibetan-style painted carvings. I am taken by surprise when I discover that this building houses the health post. When I ask the lama if it was possible to take a look he explains that he would be pleased to introduce me to his daughter, who runs the health post.

The lama's daughter greets me with a smile and in perfect English. The latter facilitates communication between us and I am able to get answers to many things I want to know about daily life in Bhijer, about

Health post in Bhijer

working conditions, equipment, the number of staff and more. Given that this area is one of the most isolated in the world, you will understand how baffled I am to find out that she has some knowledge about autism. She tells me that one of her teachers at the School of Nursing had been to a course offered by *Autism Care Nepal* for health personnel and that this teacher had told them a lot about the condition. She starts laughing when I stare at her; deadly puzzled.

My inquiries about the family with the sick boy bring about a deep sigh and a dark cloud of sadness comes down upon her. Her reaction illustrates the seriousness, which she explains to me, 'The boy is very sick. It would be best to remove the tonsils at once, but we no longer have a physician in Bhijer. The government withdrew him last year. The French couple offered to pay a doctor's wage but then we were told by the officials that we will be classified as an enterprise and thus forced to pay taxes to the state. This, however, means that the sick people have to pay money for treatments, but nobody has money here. The only thing I could do for the little boy was giving him antibiotics. Let's hope this will help'.

A mixture of sadness and anger takes hold of me; anger dominates. What happens to the 500 USD I paid for the permit? Where have all the 500 dollar 'contributions' tourists have paid since 1992 disappeared to? There is much talk about needing and using the money for development projects in this area, yet, I cannot see any. Probably, the dollars get sucked into the same vacuum-filled black holes that the money for Mustang vanishes into. Internally, I am still in a turmoil when we walk over to the lama's house for a cup of tea.

When visiting the health post I had discovered a highly unusual room for a place like this, a prayer room, and am immensely curious to hear why such a room with Buddha statues, butter lamps, incense sticks and prayer wheels is there. At first glance, it seems to be rather unusual equipment for a health post. But maybe it is not? In many societies, traditional healing methods are often closely connected to the spiritual elements of a religion. A fact that has been belittled or ignored by the Western medicine that takes over. In the remoteness of the Himalaya, the knowledge about ancient healing methods is still vividly alive and older people still rely more on an amchi when sick. The lama and his daughter do not compete because of a different understanding of diseases – quite the opposite. They are aware of the limitations their treatments are subjected to and thus combine their methods to help the people in the best way they can.

DAY 105
BHIJER – PHO
No water

It is a 1,000-metre ascent to the pass, but the good track makes me easily find a suitable rhythm. The talk with the nurse yesterday left me stirred up and now, while plodding up the hill, I wonder how the boy is today. What will happen in case the antibiotics do not help? The nearest hospital is in Dunai – many days away from Bhijer.

From the highest point, the trail zigzags down into the valley and, already from this distance, we can make out some basic huts. By the look of it, people live there. 'Let's see if they have got something to sell', Temba exclaims, heading for the houses. Of course, the people there have spotted us as well and have already gathered round by the time we reach the first hut. Tourists are not a common sight north of Bhijer because trekking groups usually turn east to visit Saldang, which is situated along one of the main trade routes to Tibet. We are way off the beaten track. However, soon we learn that I am the second tourist to pass their houses within a week!

For the locals, this is more than exceptional, but we know who the man, carrying a large, heavy rucksack, was: Nicolas. He also wanted to stock up with provisions at the shepherds' huts. Yet, the small fields further down planted with barley, buckwheat, potatoes and radish hardly yield enough food to feed the families. They have rarely anything left to sell. Even if they did, what would they do with the money? There are hardly

Shepherd

any shops in Dolpo and due to the high costs of transportation every-thing is extremely expensive. For sure, it is better to remain independent to the greatest degree possible. We consider ourselves lucky to get some freshly prepared butter, and it feels like heaven munching a chapatti with a thick layer on top.

From the kharka, the trail runs down into the deep and narrow gorge of the Tara Khola. In some places, this path has been carved from the cliff face, and though it has had some extensive maintenance, it is still steep and slippery. All of a sudden, fear grips my heart, but I have no idea why. The foreboding of disaster becomes so tangible and present that I am no longer able to analyse the situation with common sense. All I can think is that, 'Something terrible is going to happen'. It takes time to understand where this deep worry comes from. Many years ago, I witnessed three of our pack animals plunging to their deaths in a similar terrain in the Indian Himalaya. The memory of this accident has returned right now and led to this deep fear of repetition but this time a human being would be involved. In despair, I turn to look for Jomma and Mingma, walking behind me, and for Temba further down. Where is Pimba? I have not seen him for a while. My growing concern for him dominates my thoughts. Discovering the cause of these feelings of fear, however, does not help me return to a state where common sense rules and I fully expect to hear Pimba shouting for help, or to see him lying on rocks below the trail. It is an outright absurd situation but, when Pimba's laughter finally reaches my ears, I almost cry with relief.

Several hours have passed since we left the shepherds' huts, and our water bottles are empty. All the way down to the bottom of the canyon, we can hear the roaring sound of the Tara Khola and the gentle babble of brooks, but water is out of reach. We comfort ourselves with the idea of having access to water when getting to the bottom of the canyon and cross the river. However, on arrival at the bridge, we understand that the brown-grey water is, unquestionably, not good enough for drinking.

My throat is dry and hurts, and dehydration affects me physically. I remember the day when Temba and I had walked to Kangchenjunga Base Camp and back to Khangpachen, an entire day without anything to drink and I try to motivate myself, 'You did survive then and, therefore, you will survive today'. It does not have much effect. Then, it was cold, and we wandered downhill most of the time. Today, it is warm, and the long and steep ascent to Pho lies ahead.

Temba is not only equipped with an unusual ability to find trails but also to discover wells or rivulets, unfortunately, today his talent fails.

Maybe there is no water? Despondently, my eyes search the rocky slope for signs of green; grass or bushes indicate the presence of water. Nothing. I stomp on, hoping for a change round the next bend but only ever finding more stones and rocks and then, finally, the babble of a creek! There is no doubt! Anticipating a full water bottle in thirty seconds time, I start to run – only to find Temba waiting for me. Smiling sadly, he tells me about a large flat rock he had found further up the slope, which he suspects to be the local burial site where corpses are defleshed or simply left there for the vultures. The water may not be safe, and it seems better not to take any risk. We walk on.

On arriving at the first houses of Pho, we are totally dehydrated and exhausted, craving something to drink. Even so, we decide to walk past the settlement to make sure that the water from the creek is safe for drinking.

DAY 106
PHO – PUNG KHARKA
Lost in the clouds

The Great Himalaya Trail guidebook recommends that trekkers take a local guide from Pho, if they have not already employed one. Trail finding is difficult, and almost impossible in miserable weather. There are hardly any markers to guide a traveller and, in addition, the route to Mugu is unsuitable for pack animals and, therefore, rarely used by locals.

Yesterday in the evening, Temba had a long talk with some of the men from the village asking for detailed directions for the following days. 'No problem', he informs me, 'we will manage that section on our own'. Well, if he says so. I know that he would never put us in a difficult situation by professing that things are going to be easy. He radiates optimism and confidence, as usual. Great, let's go! We set out from Pho without a local guide.

The 1,500-metre ascent to Nyingma Gyanzen is demanding as the biting wind and the cold rain deprive my body of its heat. Under my rain jacket, I wear a woollen jumper and a down vest, but this does not prevent me from starting to shiver. Even in the middle of summer, snowfall can occur at any time at higher altitudes. I long for the heavy down jacket that I sent back to Kathmandu and wonder if it had been such a good idea to set out across Upper Dolpo without it. While searching for my gloves, the others are forging further and further ahead,

Clouds and fog when climbing up Nyingma Gyanzen

as usual. Since I thoroughly enjoy plodding up passes on my own, it hardly ever bothers me to find myself far behind. Occasionally, the group vanishes in the whiteness of clouds and fog, popping up again in a different location a couple of minutes later. It is almost as if a conjurer makes them disappear and turn up again.

On reaching the first ridge, marked by an old dilapidated chorten, we see the trail running through rocky and gravelly terrain along the sharp, craggy crest. Apart from the sound of furiously flapping prayer flags, there is nothing else to be heard. The place feels both wild and discomforting and I begin to understand why the locals make offerings to deities and spirits; it seems natural to believe in their existence in the remoteness of Upper Dolpo. This area is a perfect setting to shoot a film about people getting lost in the mountains.

I can still see the others on and off, resembling ghosts that come and go through white walls. There is no reason to be worried. In good weather, this walk along the ridge, which is situated in the very centre of the Himalaya range, is said to be like a walk along the spine of the planet; Kanjiroba Himal to the south and the high peaks of the Tibetan mountains to the north. Maybe I will see them next time... right now, I can only see thick clouds.

All of a sudden, I find myself in front of a vertical drop, and it is then that I realise that I am no longer following the right trail and that I have not seen my companions for a while. For how long? Five minutes?

Ten minutes? Walking back to the main track is the only sensible thing to do in a situation like this but, with the clouds and fog becoming denser, I am no longer sure where the path is. Sand, stones and rocks do not give me any indication of the right direction. In the mist and clouds everything looks the same – everywhere. Thus, staying where I am makes sense; the only sense. I start shouting HELLOOOO! No reply. What if my group has already crossed the pass and is on the way down by now? They would never hear me from the other side. I scream again and yell and scream… A boulder provides enough protection against the stormy wind and the rain to sit down and untangle my thoughts and worries.

The rain has turned into snow and the gusty wind into a fierce storm. I am freezing and, for the first and only time on my journey across Nepal, I regret being in the middle of the Himalaya instead of savouring a good meal and a glass of red wine in a cosy restaurant somewhere in Tuscany. Why the hell am I here? When the storm takes a quick break, I hear Pimba's voice shouting for me. Out of fear that my yells could remain unnoticed, I put all my energy into a loud, desperate shout. One minute later he stands beside me and together, we struggle up the zigzagging path to the pass.

DAY 107
PUNG KHARKA – CHYANDI KHOLA CAMP
Time

Good times and bad times
Steal time and gain time
Lose time and win time
Waste time and have time

Time starts and time ends
Time comes and time goes
Time gives and time takes
Time heals

Time rules the lives of 'modern' people, including me. Even after having been on the go for more than 100 days, I think in minutes and hours. It is ridiculous. I wonder how many times I have asked Temba 'How long does it take to reach X?' or 'How many hours is it to Y?' I suppose there had been hundreds of stupid questions like these. Is it of any importance if we reach a certain pass in two or four hours? Is it of any importance whether we stay at the pasture called A or the one called B and whether we need four or eight hours to arrive there? The immense emptiness and remoteness of Upper Dolpo puts time into perspective; it becomes irrelevant.

My thoughts evoke memories of my first trip to Nepal. Then, more than twenty five years ago, I had accompanied my guide Ram Bahadur Lama to his home village. A couple of days later, we had returned to the road to catch a bus to Kathmandu. Several people were sitting there, also waiting to travel to the capital. 'When will the bus come?' I asked Ram. His answer 'today' had taken me by surprise because I had thought about the exact time and so I asked again. 'Today', had been Ram's reply, which had almost driven me to the edge because I realised that he had not understood the meaning of my question. Had there been a meaning, after all?

According to information Temba gathered in Pho, local people have established a new route to Mugu because the old one was prone to falling rocks and, therefore, considered to be dangerous. Yet we do not know where the trail leaves the bottom of the valley and we cannot find any footprints to provide clues. Time passes... one hour, two hours... three? My watch no longer works, but I am too lazy to take off my rucksack and grope for my mobile to check the time. I perceive being 'without time' as being naked or being without orientation. For some unknown reason, I just have to know how long we have been searching for the trail. However, what would be achieved by that? Well, nothing, I suppose. I let it go.

Finally, a chorten comes into view, and we expect to cross the trail to Yala La there – heading to the last but one 5,000-metre pass on my walk across Nepal.

DAY 108
CHYANDI KHOLA CAMP – TAKLA KHOLA
Embarrassing performance

A breeze drives the early morning fog up the valley and drizzle escorts us the first hour after breakfast. Yet, we are lucky, and soon rays of the sun fight their way through a cloud layer that becomes thinner and thinner until, finally, the magnificence of the mountains unfolds all around us.

Today is a day full of excitement because the trail beside the river had been washed away in many places. Since bridges have vanished, leaving no means of crossing the river, I look grimly at the other bank to where an onwards track is visible. We dismiss the idea of wading through the water because the current is too strong and we are sure that the surging flow would pull us off balance. In some places, we have to traverse wet and slippery slabs, and it dawns upon me why the author of the guide-book recommends taking along a rope. We do not have a rope, but it is obviously only me who considers the traverses dangerous. The others do not even bother taking out their mountain boots but cross the slabs wearing their plastic sandals. Anxiously holding on to twigs and roots, I edge along the track, avoiding looking down at the river, rushing along with frightening power.

The traverses are not the only reason for turning me into a sad figure. In the afternoon, we have to cross some wild rivers and creeks, but there are only makeshift bridges made of old, dry branches and twigs, tree trunks or rotten, wooden planks. Whereas Temba, Pimba, Mingma and Jomma show a particular talent for high wire dancing, I peer desperately down at the raging water, but have no choice but to follow them. Usually, my 'circus performance' starts with a series of committing steps on the rickety construction, but soon the courage leaves me, and I begin to crawl on my hands and knees. In my opinion, this is more than exciting enough!

Several hours later, we cross a log bridge and climb a steep switchback trail up a shallow gully to a ridgeline. A final, slippery descent takes us down to another river; a wild and wide one. The brown water seems to boil as it careers down at a frightening speed. Waves, bouncing back from massive rocks, create whirlpools, reminding me of an oversized washing machine. Everything will be crushed and carried away by the sheer force of the river. Searching for a place to cross we walk upstream, but the only thing we find is a fallen tree trunk. I fear for the worst.

Left: Bridge in the Chyandi Khola valley **Right:** Riding over the river

In Temba's opinion it is best to take our shoes off in order to gain better friction on the rough bark. This sounds sensible, but the bark is missing in some places and there the surface is as smooth and unpredictable as ice. Taking my first, hesitant, steps, my pulse rate doubles, and I opt for an honest retreat. Another attempt follows – back again. After some to and fro, I consider it best to sit down and 'ride' the trunk across the river – as elegantly as possible, of course. Even this sort of river crossing turns out to be dangerous. The surging current drags fiercely at my feet and almost rips me off the trunk. 'Do not look at the water!' I tell myself, repeatedly, while trying to keep my eyes from wandering back to the swirls around me. On arriving at the other side, I am soaked to the bones. My group greets me with roaring applause and we can hardly stop laughing. This was probably the most pathetic show they have ever seen.

DAY 109
TAKLA KHOLA – THAJUCHAUR
The last 5,000-metre pass

If there is one dish I do not enjoy for breakfast, it is rice pudding. I know that we have almost run out of supplies. There is only some rice and a small package of lentils left. I also know that my group will do anything to make me feel good. Rice pudding in the morning, however, does not make me feel good. This is a fact. Its energy value is low and my stomach starts rumbling only an hour later. Unfortunately, I have not walked more than one third of the distance up to the Chyargola Bhanjyang by then. We had been crossing the previous passes in miserable weather and poor visibility. The bright sunshine accompanying us today evokes hopes of a perfect day and thus a chance of taking many good pictures. However, I feel tired and weak and seem to make no progress at all. Pondering about possible explanations for my weakness, I finally suspect my leg muscles of having gone on holiday without giving notice.

The pass does not come any closer and seems to grin gleefully down at me, a tiny, unimportant creature on the way to Mugu. All of a sudden, panic strikes me. What if I do not manage? I remember the worries that ruled my thoughts at the beginning of my trip. Today, they are no longer worries, circling round in my head, but are a plain, discomforting reality. The Chyargola Bhanjyang is the tenth and last pass over 5,000 metres in this region and, at the same time, the final 5,000-metre pass on my journey across the Himalaya. This very moment, I stare despondently at the saddle high above me, desperate, exhausted and tired. I am totally done, 'You achieve anything you want – if only you put all your heart and soul into it', I tell myself over, over and over again. I visualize *Mission Impossible I* and *II*, when I succeeded against all odds and, as if in a slow motion movie, I start moving, placing one foot in front of the other. I avoid looking at the pass, knowing that seeing the barely decreasing distance would crush me totally. My mind starts counting the steps: one, two, three… until I reach fifty and stop, take a short break and begin to trudge up the barren scree gully once more. I do not know how long or often I repeat this procedure. Any conception of time has left me when I finally arrive at the chortens that mark the highest point. The sun disappeared behind gloomy clouds some time ago and violent gusts of biting wind tear at the gauzy Tibetan prayer flags that decorate the chortens. Even so, I grab my camera from my rucksack to take a short video of this 'historic' event.

From Chyargola Bhanjyang I look down into a different world; a world of green pastures and trees. Assuming that the group has already reached the valley by now, I am sure to find some hot tea and a portion of rice waiting for me when getting there. I start scrambling down the slope.

DAY 110
THAJUCHAUR – TIYAR
The new bridge and neighbourly help from China

One or two years ago, trekkers had to walk uphill towards Mugu to cross the river safely via a suspension bridge. But before reaching the Mugu Khola, local people tell us about a new bridge further down the main trail. How lucky we are! This gives us some extra time, and we look forward to a long break in the warm mid-day sun. Near the confluence of the Mugu Khola and the Cham Khola there is a shop which does not only sell the main staples (rice, lentils, flour and beer), but also some unique luxuries (fresh vegetables, butter and cookies). Our lunch turns into a banquet!

Why is there a well-stocked shop in the middle of the wilderness? The question is easy to answer, because of the Chinese.

Apart from the Kodari (or Araniko) Highway, which was built in the '60s, there have been a couple of new road construction projects in the last ten to fifteen years financed by the Chinese. Today, there are five new roads which – coming from China – end somewhere in the high mountains of Nepal; many trekking days away from the Nepali road network.

Banquet beside the river

In 2008, China started to extend the railway network in Tibet, and one day in the (near?) future, trains will run all the way from Lhasa to the Nepali border. This project raises the question, why does China invest a fortune in an almost 2,000-kilometre long expansion of public transport running through a sparsely inhabited region and ending at a border when there is no rail network in Nepal.

Neighbourly help or invasion attempt? Selflessness or perfectly planned control function? Whenever and wherever I asked people along the way, I received conflicting replies.

Here are some facts:
- The Nepalese government uses the close relationship with China and the financial support to offset the Indian influence.
- After the violent suppression of protests in Lhasa, Tibet, in 2008, Nepalese authorities began to prohibit cultural and religious meetings, closely directed by Chinese officials. Nepalese officials say that as activists prepared to mark the anniversary of the 1959 Tibetan uprising on March 10th, and of the 2008 Lhasa riots on March 14th, a delegation of Chinese intelligence officers arrived to oversee the suppression of protests.
- According to Nepalese officials, the Chinese are about to establish a series of concentric 'security rings' around Lhasa. The outermost will run through Nepal, partly based on Chinese networks, operating under the cover of NGOs, language institutes and small businesses.

- Chinese agents have infiltrated the Tibetan community in Nepal, which numbers about 18,000 people.
- According to a trusted U.S. embassy cable published by WikiLeaks in 2010, 'Beijing has asked Kathmandu to step up patrols... and make it more difficult for Tibetans to enter Nepal'.
- Vital escape routes have been choked off by the Chinese.
- Robbie Barnett, director of Columbia University's Modern Tibetan Studies Program says 'In many respects, China now determines Nepal's local and foreign policy'.
- For years, Nepal relied heavily on aid, trade and investment from India. In return for its co-operation with the Chinese, the impoverished nation of thirty million people has found a new benefactor. When visiting Kathmandu in March last year, China's army chief Chen Bingde pledged twenty million USD in military aid. This year, dubbed by Beijing a 'year of friendly exchanges' between China and Nepal, began with a visit by the Chinese Premier Wen Jiabao. He offered 119 million USD in aid.
- A long-standing agreement under the UN High Commissioner for Refugees, allowing Tibetan refugees to cross Nepal on their way India, is still in operation, but numbers have declined recently.

DAY 111
TIYAR – LHUMSA
Do I have two heads or three noses?

Crossing Dolpo had been demanding, and so I appreciate the easy stages down to Gamgadhi. It is almost as if all the Himalayan gods have created this trail for me only; a reward for my endurance and a compensation for the hardships. Today, the trail runs beside the river, and there are hardly any ups and downs. I totally enjoy my life as a trekker.

The people living in the Mugu area are different from those in Dolpo. There, we were always approached by cheerfully laughing or smiling people and though they had been exceedingly curious about everything, they showed dignity and respect for privacy. It did not happen once that a *Namaste* or *Tashi Delek* was ignored or not answered and, usually, any encounter ended with a conversation accompanied by a lot of facial expressions and gestures.

The population of Mugu (50,000 inhabitants) consists of Buddhists (approximately thirteen per cent) and Hindus. Droughts and famines

Foreigners and strangers are met with scepticism

have afflicted the region repeatedly and, as a result, the locals are severely poverty stricken. The general educational level is appallingly low; eighty per cent of the population is illiterate and only a small percentage speaks and understands Nepali. Those who do are men because their schooling is given priority. Like in many other rural areas, boys attend school, whereas their sisters have to stay at home. Thus, women do not have access to information, given out by the government in Nepali only.

Only about fifty per cent of the population has access to safe drinking water. Infant mortality is extremely high, even by Nepalese standards, and life expectancy is low – around thirty five years for women and forty five years for men. In Mugu, everybody is poor, but those belonging to the caste of Dalits (untouchables) are even poorer, like anywhere in the country. The traditional Hindu practice is based on social caste discrimination, which entails that Dalits are often denied access to water resources in the villages. The poorest of the poor, however, are the Dalit women. They are alienated on the basis of class, caste and gender and, therefore, suffer from triple discrimination: oppressed by high caste people (equally affecting both male and female Dalits); oppressed by the Hindu patriarchal system and by Dalit men. Ninety per cent of the female Dalits in Nepal live a life below the poverty line and eighty per cent are illiterate. These women are exposed to serious health issues, sex trafficking, domestic violence, and they suffer from social, political, and economic power-lessness. Dalit women are the 'invisible citizens' of Nepal, who have to wait at the public water taps and wells until so-called high caste women are finished. If they interfere, they become victims of violence and mistreatment.

A woman explains the situation, 'We face death even from a common and curable disease because we don't have access to medical treatment here. Sometimes we even feel like we are living in a desert when we face scarcity of water'. Many of them drink the washing water after having cleaned their husbands' hands and feet with it. How is this, given that Nepal holds second place in the world when referring to water resources?

Although the custom to send heavily pregnant women to dirty and filthy cow sheds is based on tradition, it clearly shows their subordinate role in society. Nobody seems to care that this tradition takes its toll regularly. Each and every year, women die due to infections, internal bleedings and other fatal complications while and after giving birth. Rapes are the daily routine for girls and women because being 'untouchable' does not include sexual abuse. Men from higher castes have the 'right' to take what they want – whenever they want. Hinduism turns a blind eye on this situation.

Today, it is common knowledge that any kind of improvement or development within a society can only be implemented by women and, therefore, a couple of international organizations support Dalit women in their fight for their human rights.

Apart from the Dalits, mainly Khasas live in Mugu. Originally, they had been followers of Tibetan Buddhism, but became converted by Indian Hindus who settled in western Nepal (between the ninth to fourteen centuries) because of the Islamic invasion. The majority of the Khasas consider being called Khas humiliating and, therefore, they call themselves Chhetri or Thakuri.

Even decades after the official end of the caste system, most Hindi people are aware of their status, related to a certain caste, and their lives are still characterized by strict rules that originate in tradition and religion. Despite this, poverty afflicts each and everybody in Mugu regardless of caste or social status and, as a result, the line dividing Dalits from other groups has begun to disappear in some places. Occasionally, people from higher castes sit together with Dalits and even eat together.

Yet, even with times changing slowly, the inhabitants of this region are firmly in the grip of traditions, religion, caste system, superstition and ignorance imposed by the government that does not provide education. As a result, tourists are met with scepticism.

Wherever I walk, stop or sit down, people stare at me as if I have just arrived from a faraway galaxy. Do I have two heads or three noses? It almost feels like both. Nobody answers my greetings and I hardly meet anybody who smiles at me. In the beginning, the thought of something being wrong with my clothes haunts me and nervously I check whether all buttons are closed and whether my skirt reaches down to the ankles. Everything is perfect. It is virtually impossible for me to get used to the stares and I begin to miss the Tibetan culture with its friendly smiling people.

DAY 112
LHUMSA – GAMGADHI
Back to civilisation

Gamgadhi is not far away from the camp, and I had hoped to be able to enjoy a late breakfast for a change. My group, however, had another idea about how to interpret the word 'late', and so we are back on the trail only half an hour later than usual. According to the map, it will be an easy walk along the river. Like yesterday, the well-maintained path runs through a beautiful forested area and we pass a number of hamlets where locals sell sweet peaches. After the basic diet of Upper Dolpo, biting into a fruit with droplets of juice running down the chin feels like heaven. My group takes pleasure in this luxurious life too, and so I wonder how many locals will not need to go to the market to sell their fruit. I think that we consume ten peaches each between Lhumsa and Gamgadhi; at least.

The closer we come to the small town – that has an airstrip – the more people we meet; children on their way to school, young students, women carrying fruit and vegetables to the local market, tradesmen, grandmothers leading their grandchildren and young mothers carrying their babies. I notice that many of the latter are hardly more than children themselves. It is common knowledge that child marriages are still widely practiced in rural parts of Nepal and rough estimations indicate that about eighty per cent of girls get married between the age of ten and nineteen. Poverty, tradition, culture and religion are the main reasons for early marriages. The parents cannot be blamed as they only want the best for their daughters. They do not know better.

In Nepal, girls have, in general, a lower social status than boys. Poor families often consider girls a burden because they will never contribute financially or take on the role of a breadwinner. If a girl is still unmarried by the age of eighteen or twenty, she is met with scepticism and rejection, both by her parents and society. Additionally, she becomes a financial risk for her parents, who have to offer a much bigger dowry to the parents-in-law the older she is. This may result in bankruptcy.

Enlightenment equals salvation, to put it simply. Reaching this state means an end to suffering and ignorance, and thus it is a worthwhile

Gamgadhi

goal for followers of both Hinduism and Buddhism. Many people still hold the (erroneous) belief that a girl who gets married before her first menstruation will enhance their parents' or grandparents' chances to achieve enlightenment after death.

I am looking forward to experiencing the luxuries that civilization usually guarantees – but there are surprises in store for me. What I had expected to be an easy and short stroll up the hill turns out to be hard work in the hot and humid air of lower altitude. It does not take long until I am drenched in sweat, and I am already picturing myself standing in the bathroom under a never-ending stream of hot water from the shower head when the first houses of Gamgadhi come into sight.

Temba asks for the best hotel in town, but 'best' proves to be relative. There is not even a separate bathroom where I can wash myself. The only luxury is a toilet. Luckily, Mother Nature is on my side and solves my problem. Only half an hour after we have moved in, the heavy drops of torrential monsoon rain are pattering down. I step onto the rickety, wooden balcony on the backside of the hotel building and enjoy the shower I was longing for – fully dressed, of course.

Tomorrow, Pimba, Mingma and Jomma will leave for Kathmandu. It is a five-day journey, entailing a three-day walk to Jumla and a two-day bus ride from there to the capital. Of course, their departure is a good enough reason for a chang party. I will miss them and their joyfulness, and am already looking forward to meeting them again in Kathmandu.

TEN

JUMLA/HUMLA

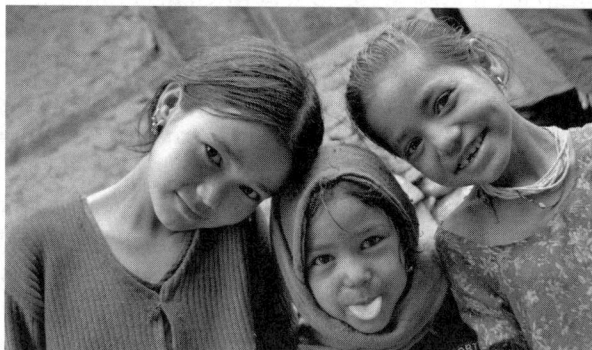

Girls from Yari valley

DAY 113
GAMGADHI – BAM
The unknown enemy's return

The unidentified enemy that attacked my hand many weeks ago seems to have relatives here in Gamgadhi. Yesterday, while writing mails and a new blog entry, I was bitten in my left foot, and only two hours later clear signs of an infection appeared: inflammation, swelling and a throbbing pain. I hardly made it to the toilet that is not more than five metres from my room.

'A good night's sleep does wonders', I thought and went to bed early. Sleep? No way. By the time I had finished listening to the third CD on my iPod, the pain had already spread to my thigh, and I decided to consult my First Aid box for antibiotics and painkillers.

Today, the acute pain is gone, and the foot is no longer hot, but it is even more bloated than yesterday, resembling more a shapeless lump of flesh than a foot. When trying to force it into my trekking boot, the stabbing pain makes me change my mind on the spot. Maybe I should ask a doctor to take a look at it? My plan was to go to the local hospital anyway to hand over some information leaflets about autism and, depending on their interest, to hold a short lecture. On arriving at the hospital and seeing the littered property, I begin to doubt that consulting a doctor here is such a brilliant idea. The grimy visitors' chairs make me cringe and change my plan for good. What else than antibiotics and pain killers could they give me? Those, I have enough of.

A foot that does not fit into a proper shoe is not ideal for walking. I am aware of this, but I do not want to stay in Gamgadhi either. 'Do as the locals do', I think, while applying disinfection to gauze and wrapping it around my foot. Ten minutes later, Temba and I leave the town, me, wearing a pair of plastic sandals, just as the Nepalese locals do.

Later that day, I regret having decided to continue a couple of times because my foot hurts whenever the trail runs up or down – and flat sections are rare in Nepal. In addition, the current heat and humidity of monsoon time affects me and the combination of pain, antibiotics, physical exercise and climate does not contribute to my well-being. Waves of tiredness, despondency and gloom keep coming and going and I do not know how we manage to get to Bam, but we do, finally.

DAY 114
BAM – RIMI
Karnali River Zone

We have, finally, arrived in monsoon country. Sometimes, I wake up in the middle of the night to the sound of rain, clattering softly on the tent and think about how it's going to be to pack the wet things in the morning with the rain still pouring down. Yet, usually, we are lucky because, by the time we have to leave, the rain has stopped. Last night, heavy rain hammered down with vigour, hitting my tent in torrents, and it is still pouring down when we set out for the next stage. I do not fancy sunshine in general since sunshine means heat, but today is different. We hope for the sun to break its way through the clouds to dry the damp sleeping bags and the tent.

The nightly rain has turned the track into a muddy creek and, since my foot is still swollen, I wade through ankle-deep mud in my plastic sandals. As long as it is 'normal' dirt, I do not care, but in the villages it is not just 'normal' dirt, but a combination of everything humans and animals leave behind. Apparently, Temba and I are the only people who feel offended by the muddy main streets. The local people walk around with bare feet, whereas we make committing leaps from stone to stone, trying to avoid stepping into the muck.

This area is a part of the Karnali River Zone, which is not only one of the poorest and most isolated regions in Nepal but in the world. The striking poverty has been worsened by the Maoists raging in this area during the civil war, and even though the fighting is over, the local population still faces serious problems. The changes they need are related to hygiene and agricultural technology, the latter to increase the yield and thus improve their nutritional status, which is poor. The lack of safe drinking water tops the list of the most urgently needed improvements because many of the other difficulties the people deal with derive from this lack. Access to clean water sources will bring about a change in peoples' lives – people who currently have to use contaminated water from irrigation canals, open pools and rivers for drinking. They have no choice. Sick children cannot attend school – and without education there is no chance for a better life. Sick adults cannot work in the fields – without crop yields there is no income. A vicious circle that has no way out.

We walk through villages and settlements where the, apparently, helpless and desperate inhabitants squat in front of their houses, waiting for something to happen. However, nothing is going to happen unless they start doing something themselves. The atmosphere is rife with gloom and despondency, and I begin to understand that the harrowing poverty I meet here is the norm in Nepal. The Nepalese government does not seem to be too eager to provide help and education. Educated citizens are critical citizens, who may question or challenge decisions related to the distribution of foreign aid money.

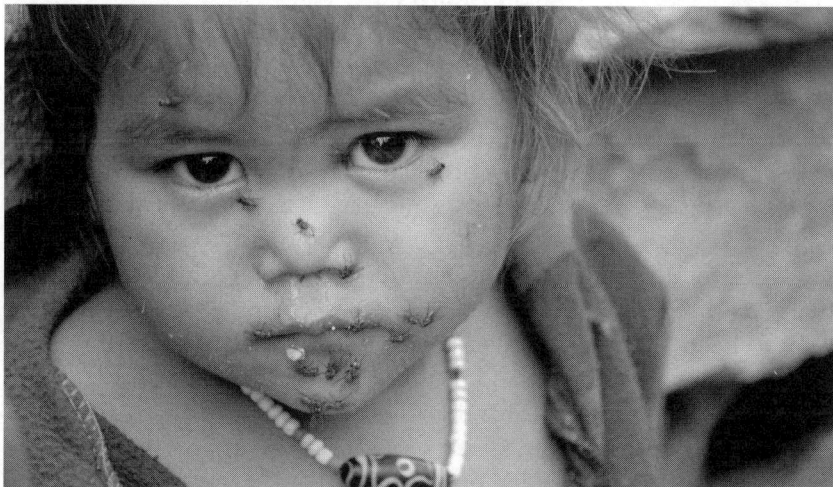

Vicious circle of poverty

The Associated Press:

'The UN's new co-operative agreement with Nepal over the next five years will amount to some 700 million dollars. That adds up to an estimated ten billion dollars in aid money that Nepal has received since the late 1950s'.

'When we go to any government or local administration offices in the villages or even in the cities, it is usually run by the so-called elites', a student says. 'They have always treated us like servants and they are the masters',

Published May 31st 2012 Associated Press

It is also here that I see the first and only people on my trip affected by leprosy, a disease that is related to poor conditions such as inadequate bedding, contaminated water and a poor diet.

In one of the villages, the inhabitants mistake me for an NGO representative and urge all children they can get hold of in front of my camera lens. The majority are dirty and poorly dressed, show signs of malnutrition and have, at least, one infected wound. When it turns out that I am no more than a normal tourist on the way to Simikot and without any money to give out, the mood of the local men changes abruptly. They want money for the photos. A heated discussion starts, and in the end I delete all the pictures I had taken. Disappointed and angry I leave the village behind.

DAY 115
RIMI – SARKEGHAT
Alternative route II

Today, I decide to give my mountain boots a try as there are two passes ahead and the terrain will be difficult. As long as the trail runs through flat areas, I do not have any problems walking with sandals like the locals, but as soon as I have to walk longer sections up or downhill, I prefer the support a proper shoe provides. When forcing my left foot into the boot, things look and feel good – but not for long. Less than one hour later I regret my decision as the growing pain tells me that there is no way I can continue in the boots. What to do? Is it sensible to walk over a pass wearing sandals and risking a strained or sprained ankle? This would even be more disastrous than what I have to cope with right now.

According to the map, there is an alternative trail from Piplan to Simikot along the river. This route is longer but easier because we can avoid one of the high passes. This sounds promising and Temba collects information about where to eat and sleep. There are many villages on the way, and since this trail is the main route to Simikot, we should not have any problems finding food or shelter. My hurting foot screams with joy when I release it from its mountain boot prison. Changing the route is the only sensible decision I can make in this situation. Anything else would be stupidity. After a short break for tea and biscuits, we head westward from Piplan. The path could not be better and sandals are ideal for wandering along the flat, well-maintained trail.

Today's destination is Sarkeghat, a large village comprising of a few hotels, shops and restaurants. Unfortunately, the standard of the hotels is poor, and none of the places appeals to me in particular. Of course, we always have an emergency plan and Temba pitches my tent on the roof of a building. While he sets up my red 'home sweet home' I search the hotel for a bathroom and a toilet, but to my despair, it is lacking such amenities. There is not even a water tap. My heart sinks when I realise that my dream of a refreshing shower after a long day will, in all likelihood, remain a dream.

The hotel owner explains that the only water tap they have in the village is on the other side of the river, beside the police station. This information raises new hopes, but presents a problem; how do I wash in a public place? Finally, I opt for a 'Wetness Cabaret*' performance and take a 'shower' at the village water tap, after darkness has set in, fully dressed, of course, like the local women.

DAY 116
SARKEGHAT – BRIDGE LODGE
'I am hungry'

Yesterday's dinner was a disaster: overcooked rice, bland soup and no vegetables. I rarely complain about food, but last night I did. Probably, the cook was thoroughly dismayed and tries to make up for it now, the breakfast is delicious! My good mood returns immediately and the cook receives my compliments on the food.

There is a police check post in Sarkeghat, but since the policemen focus on their breakfast, none of them are interested in my permits. That's perfect as the given dates are not correct and this would, doubtlessly, result in

---

* Wetness Cabarets were once hugely popular in India. Since striptease was not allowed at that time, young women wearing thin, wet saris danced in night clubs.

endless discussions and, probably, serious problems. Over time, Temba has developed a remarkable skill in dealing with the police and talking me out of trouble, but it is better this way; no questions – no excuses.

When travelling alone with Temba, I usually send him ahead to order lunch or dinner in the next village. It takes at least an hour to prepare Dhal Bhat, and I consider it pure luxury to arrive at a lodge or restaurant where a plate of steaming hot food is waiting for me.

Temba is already on his way to the hotel at the bridge locals had recommended, when a small group of young men catches up with me. As expected, they stare and gawk at me without saying a word. My umbrella is useful in a situation like this one because it allows me to disappear. My 'theory' in these situations is that if I cannot see anybody nobody can see me. Usually, men lose interest quickly, which is all I want.

'I am hungry', one of the young men stammers when arriving beside me. 'I am hungry too', I reply since I do not want to be impolite. I step to the side to let him pass. 'Give me money', is his next sentence. Do I understand correctly? In the most friendly and calm tone I tell him to ask my guide for money because it is him who carries the wallet. At the same time, I think of the bundle of banknotes in my computer bag and the stories about the unresolved disappearance of a wealthy merchant local people had told us a few days before. The young man hesitates a moment, exchanges a few words with his friends, and together they continue their journey.

Outside the hotel at the bridge, we meet again. Temba and I invite him for dinner, but all he wants is money, probably to buy alcohol. We are not going to support this.

DAY 117
BRIDGE LODGE – KHARPUNATH
Toni Hagen

We continue our journey towards Simikot, and I enjoy the morning walk and the cool, fresh air that blows up to me from the river, careering through the valley at a frightening speed and with intimidating force. The trail is impressive, the steepness of the mountain has made it necessary to cut hundreds of steps into the rock and, once again, I am surprised by the artfully-constructed paths Nepalese people have built over time without any heavy machines. How long did it take to cut out all the steps, I wonder.

Way to Kharpunath

Later, we walk through forests and agricultural areas where locals are busy cutting the weeds, and where apple trees line the trail to the villages.

We meet a lot of people traveling to Simikot who are all traders or businessmen, apart from two. One of them is the local postman, and the other one is a man in his late twenties, carrying his young wife on his back. Her face, distorted with pain, and her moans tell me that she is seriously ill, and I wonder if a complicated pregnancy or problems after having given birth cause her sufferings. I cannot help thinking of child marriages again. The status of a young bride in her husband's family depends on her fertility, and the only thing that counts is the number of children she gets; preferably boys, of course. Early pregnancies, however, are an enormous health risk for girls, resulting in higher rates of illness and death for both the mother and the baby. Among young mothers, the worldwide incidence of premature birth and low birth weight is higher and anaemia, pre-term delivery and low birth weight are more likely to occur than among mothers in their twenties. Frequently, protracted or complicated deliveries lead to internal injuries resulting in the young mothers bleeding to death.

The nearest hospital is in Simikot, and it is doubtful whether they get there today.

In 1950, the Swiss geologist Dr Toni Hagen (1917-2003) was the first foreigner who was allowed to trek throughout Nepal during his geological and geographical survey work, mapping on behalf of the United Nations. Two years later, in 1952, he was employed by the government of Nepal and worked for the United Nations again. All in all, he walked over

14,000 kilometres across Nepal. In the '70s, he published his book *Nepal*, which contains useful statistics, enabling us today to evaluate the development in different fields.

Infant mortality

'70s: 2-300/1,000	Today: 40-50/1,000	Karnali: 170/1,000

Life expectancy

'70s: 40-50 years	Today: 60 years	Karnali: 40 years

Doctor/inhabitants

'70s: 1 doctor/40,000 people	Today: 1 doctor/5,000 people	Karnali: no info

Illiteracy

'70s: more than 80%	Today: more than 50%	Karnali: 80%

The last decades brought about many improvements for the Nepalese people, but there is still a long way to go. Statistics are often misleading and can hide facts – here, for instance, they provide little more than a general overview and do not show or explain the increasing gap between urban and rural areas. The life expectancy of the urban population and people living in the Karnali River Zone, differs by approximately twenty years! Similar differences can be found in virtually every field and explains why there is such a large rural exodus – a key issue in Nepal. Life in the hills and mountains of Nepal has not changed much over time. Hygienic standards are poor, and since we left Gamgadhi, every lunch or dinner turns into a desperate but hopeless fight against swarms of flies. No matter where I sit down to eat, the pesky flies find me and the food and a fierce battle follows. The villages are filthy, and the same goes for the few hotels in this region. There is no safe water supply, and I cannot remember having seen a toilet for a long time, apart from the one in Gamgadhi.

Twenty-five years ago, the situation in the Annapurna and Everest areas was similar but education has brought about knowledge, awareness, understanding and, eventually, innovative ideas. As a result, hygiene, besides many other things, has improved considerably. Maybe I should return to Jumla in twenty five years? Would there be changes? Who knows.

DAY 118
KHARPUNATH – SIMIKOT
The end?

I enjoy the climb to Simikot. Temba and I meet many people on their way to the market and enter into all manner of entertaining conversations that make the time pass quickly. Around noon, we reach the small town, situated on a plateau high above the Humla Karnali Nadi.

We have almost run out of cash, and I ask about an ATM. None. Simikot's tiny airport is the only sign of modernity. Luckily, I always have some 'emergency' Euros in my wallet, and so all we need to do is find a bank. This should be an easy job I reckon, but reality proves me wrong. The first bank is closed and the second does not change foreign currency. The third and last bank lies hidden in one of the narrow side lanes and I would not have found it without Temba's help.

The only person working in the bank is a young man who greets us in a friendly manner. 'Of course, you can change foreign currency here, Madam. Please, give me your passport', This, however, is not possible, because my passport is in Nilam's office, in Kathmandu. He needs it when applying for visa extensions and permits. 'Is a copy sufficient?' I want to know. The bank clerk's attitude changes from 'very friendly' to 'friendly' when grabbing the paper. Nilam made the copy in March, a few days after my arrival in Nepal and, according to the dates given on the copy, the visa expired several weeks ago. While examining the document, even 'friendly' disappears from his repertoire, and in a sharp voice he exclaims, 'You are illegal here in Nepal! No money. End of discussion'.

I try to explain my situation, but this only leads to increased bureaucratic stubbornness on his side. 'Friendly' also vanishes from my repertoire. Before the conflict escalates completely, we ring Nilam's office and ask him to send a fax with the latest updates on my visa details to the bank in Simikot. Needless to say, the fax machine is out of order here. I am close to a nervous breakdown.

I hand over my 'emergency cash' to Temba, and now it is his turn to try his luck. However, by the looks of it, the clerk derives enormous pleasure from 'torturing' us. 'Your passport is not enough. I also have to see the license, approving you as a registered guide', he tells Temba. Nothing easier than that. My guide hands over the document. In Nepal there are different associations issuing certificates and, of course, Temba has got his license from one the clerk is not allowed (or willing?) to accept. For a fleeting moment, I suspect him of waiting for an unofficial increase of his wage, but I am not totally sure and let the chance go by. Our argument stops when I, finally, lose my patience and leave the bank office with an air of seething anger around me.

What to do? Is this the end of my trip across Nepal? Neither cold had managed to delay or stop me nor heat, snowstorms, monsoon rains, biting insects or wild rivers. Now, it appears that an argumentative bank teller from Simikot has put an end to my traverse. This is downright ridiculous!

Temba's calm and balance are astonishing, and once again it is him who sorts out my problems. Outside the bank, he takes out his mobile and starts phoning. Five minutes later there are no problems left. One of his brothers has some close friends in Simikot, and they will lend us as much money as we need.

Later in the afternoon, we sit with our new friends, chatting cheerfully and drinking chang. A pile of bank notes is right in front of us. Enough cash to reach the border and to return to Simikot.

DAY 119
SIMIKOT – CHYADUK
Final sprint

Simikot is my first contact with Western civilization after a long time, and I enjoy the feeling of luxury when waking up in a 'real' room with attached bathroom, perfectly equipped with a clean toilet and a shower. I treat myself to a prolonged breakfast with all sorts of dishes, ranging from Tibetan breads to vegetable curries. Now, we are ready for the last section of the journey.

The sky is grey, and the monsoon clouds hide the peaks of the surrounding mountains. Gazing at the maze of dark clouds, I wonder if the Australian group I had met yesterday will make it to Kathmandu today. They, too, had walked the Great Himalaya Trail and, needless to say, we had exchanged experiences and all sorts of news.

In bad weather, all flights are cancelled because the airport lacks a radar system, and thus take-offs and landings are virtually impossible. The rusted remains of a crashed airplane are an alarming and unpleasant reminder of the accidents happening every year in Nepal.

According to the times given on the map, we should be able to get to the border within six days, but Temba and I agree on pushing on for a five-day final sprint. After a long and rich breakfast, consisting of Tibetan breads, vegetables, pancakes and fried eggs, we set out. From Simikot, the trail runs 1,000 metres down a steep slope into the valley of the Yari River.

Working women

Admittedly, this is a good start, but I am already trying to imagine what climbing up again will be like, knowing that there will be no way of escaping the ascent. In the humidity and heat of monsoon time, a 1,000-metre climb will be hell, and the mere thought of this inevitable torture makes me feel dizzy.

This region is particularly pretty and surprises me in different ways – even without dramatic 8,000-metre peaks. Small villages line the route, and the general friendliness of the local people makes me want to stay. We walk alongside fields of flowering potato plants, ripening buckwheat and barley, and pass walnut groves and apricot plantations where the first orange-coloured fruits seem to call out 'Please, feel free to taste us'. Unfortunately, our arms are too short to reach them, but sometimes we can buy tasty apricots and peaches in one of the villages.

Carol Dunham, an anthropologist, wrote about this area:

'I think Humla is honestly one of the most culturally fascinating places in all of Nepal, a cultural tapestry woven from ancient Khasa kingdoms, ancestors of the grand Zhangzhung kingdom of the north, with a mix of Rajput and Thakuri blended into the mix'.

With the distance between us and Simikot increasing, Tibetan culture takes centre stage again. Women, wearing chubas, are busy spreading out the ears of corn to dry them on the flat roofs of houses. Colourful Buddhist prayer flags flutter lazily in the wind, chortens mark places of significance, and we see an occasional stupa. It is good to be here.

DAY 120
CHYADUK – TAPLUNG
Protecting the culture

Compared with other walks along valleys in Nepal lying behind, this section, following the Yari valley, is a Sunday afternoon walk. However, due to our plan for a 'final sprint', the days are long and I am pretty tired when we arrive at the first houses of Taplung.

Over time, I have developed an ability to evaluate the quality of schools. For me, it is sufficient to exchange three or four sentences with local children to determine whether there is a private or a government school in the village. No doubt, in Taplung there is a private one; an excellent one. The children do not only come with the standard questions about my name, country and age, but they are able to answer my questions and even tell me a few things about themselves. Soon I find out that about 150 pupils attend a boarding school here, founded by a German woman. Her project became a co-operation project with the Himalayan Children's Society (HCS) – a Humla-based organisation which based the project on the idea that the young people of Humla themselves should decide the future of their home villages. The NGO focuses on developing the educational resources in the district, including teacher training and initiatives to encourage families to enrol their children (particularly girls) at school.

The project, however, does not give centre stage to education alone. Brain drain is a pervasive problem that slows down development in many poor countries – including Nepal, which experiences the migration of highly skilled workers. Is there a way to avoid this? The Himalayan Children's Society approaches this problem in the same way Action Dolpo does, placing its main focus on preserving the local (Tibetan) culture. This will, hopefully, create a genuine and strong bond which will encourage many young graduates to return to their home areas where they can get jobs as teachers and nurses.

The boss of the boarding school has developed an impressive concept which partly concentrates on preserving the Tibetan culture and partly focuses on research with the purpose of reviving a cultural heritage that has almost disappeared. Now, the children's curriculum includes playing traditional Tibetan musical instruments and learning old folk dances and songs. In his opinion, this will result in a high regard for fundamental values, helping young people to find their own identity in a changing world, even when they one day leave their homes for good.

It is an important day for the pupils of the boarding school in Taplung.

Cultural show at the boarding school

Some officials from Nepalgunj have arrived to attend a cultural event featuring traditional Tibetan songs, music and dances. They ask Temba and me to attend the presentation and, needless to say, we are pleased to accept. The auditorium is bursting at its seams – there are not only 150 pupils and all the visitors but also the entire teaching staff, the parents, the staff from the boarding school and the villagers. All faces turn towards the stage where the young musicians begin to tune their musical instruments. The members of the dance group line up and, on a sign by the headmaster, the show starts. It does not take long before all of us are totally engrossed in a different world, the world of ancient Tibet. The audience shows immense enthusiasm and, as soon as there is a short break, thundering applause causes the whole building to shake.

DAY 121
TAPLUNG – MUCHA
Traffic jam

Everywhere in Taplung, pupils greet me with a cheerful 'Good morning. How are you today?' Since Tunket, our destination for today, is less than five hours away, we have enough time for some more chatting and several cups of tea with the boss of the boarding school. I like this place, and when we part I promise to visit the school on my way back from the border.

The trail between Simikot and Hilsa is a main trading route with heavy 'traffic'. The enormous and obviously insatiable demand for cheap Chinese products leads to crowded trails. Everywhere, we meet herdsmen busily driving hundreds of pack animals to and fro. The men whistle, shout, utter guttural cries and throw stones, and it is surprising to see this sort of communication working so perfectly. Yaks carry the heaviest loads: machine parts, kitchen stoves, furniture... Mules and horses transport boxes with bottles and cans, packages with butter or milk powder and other durable goods such as umbrellas, clothes and shoes. Goods that can be repacked in smaller packages, like rice and flour, are carried by goats. Many of them have old-fashioned bags fastened to their backs. These bags are made of hand-woven fabric and vegetable-tanned leather stitched together with woollen threads as there are neither zips nor Velcro.

The goats tend to be extremely shy and try to get around 'living obstacles' – such as me – by leaving the trail. Of course, the herdsmen are not enthusiastic about this at all and ask me to step aside; no problem. The horses and mules are frightened by my umbrella and I am prompted to close it; no problem. Yaks simply walk straight on and do not mind obstacles.

After having plodded behind a yak caravan for a while, I decide to pass it, aware of the fact that manoeuvres like this are extremely hazardous when there is a precipice beside me. Animals are unable to judge the width of their loads, protruding on both sides of their body and thus, every year, unwary trekkers have accidents or are killed after being nudged off the track by passing animals. Walking on the inside edge of the track, I pass all yaks safely and step back to the middle of the trail when an angry yak decides to overtake. At the last moment, I jump out of its way but, for some unknown reason, to the wrong side. Luckily, a bush stops my fall before I roll down over the precipice.

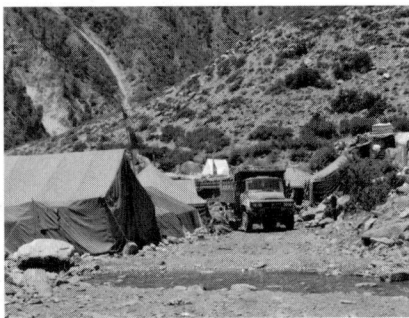

Left: Goats as pack animals **Right:** Pani Palbang

There is both a police station and a check post in Mucha, and my papers for this district are scrutinized meticulously. According to the dates given on the permit, I am no longer allowed to wander through Humla, the permit expired a week ago. Temba saves the situation (and me from paying additional fees) by deploying all his diplomatic skills and talking in a friendly – but convincing – tone with the policemen. The only words I understand are, Great Himalaya Trail – four months – Taplejung. Of course, in the beginning the policemen stare at me in disbelief, but then, admiration and respect follows. Now, it is my turn to impress them. I unpack the Notebook from the bottom of my rucksack and ask if they want to see some of the places I have passed so far. This manoeuvre to distract them works and one minute later, the entire staff of the Mugu police station gather round in front of my Notebook, looking at Sherpani Col, Namche Bazaar, Muktinath and Shey Gompa. When we set out again, they wave us a friendly good-bye and wish me good luck.

Carrying a computer across Nepal can be useful.

DAY 122
MUCHU – YARI
The road

In the early hours of the day, we pass through Tumkot, cross the river and climb out of the valley. In several places the trail is narrow and, filled with horror, I think of what had happened the day before. Whenever I meet a caravan today, I climb up the slope beside the path – as far as possible.

When Pani Palbang is in sight, we can also see the dusty dirt road coming from China. One day in the future, it will lead all the way to Simikot. And from there? What's the purpose of building a road that will end in the remoteness of the Nepalese mountains? The buying habits of the Nepalese people speak volumes, all over the country, cheap Chinese products are favoured.

Pani Palbang, consisting of more tents than houses, is the last stop for jeeps and lorries and, as a result, the tiny settlement has developed into a transhipment point over time; life is in full swing all day long.

The people unload vehicles. Bulky packaging materials are removed and end up on one of the garbage dumps one spots virtually behind every tent and beside the road. Transportation has to be planned with consummate care as every bottle that breaks entails financial loss. It is also necessary to repack goods into suitably sized loads before loading them on to pack animals. Everywhere in Pani Palbang herdsmen are busy sewing up the old-fashioned bags I saw yesterday.

While women feed their babies or do the washing in the dirty water of the nearby creek, men drink, play cards or chat away with friends. Music is blasting from radios powered by old car batteries and, at one of the tents, a camp beauty is waiting for a visitor. The people have other things to do than to pay notice to me, a passing tourist. This is highly unusual in Nepal.

We cannot avoid a rather boring walk along the dirt road, with dumped litter 'decorating' its shoulders. Somewhere between Pani Palbang and Yari, a jeep stops beside me, and the driver asks if I want or need a lift. 'No, thanks. But I will need a jeep from Hilsa tomorrow – about two o'clock in the afternoon'. I tell him. He promises to be there at the given time.

DAY 123
YARI – HILSA
The last day

Yesterday, we had been welcomed warmly by our hosts, an elderly Tibetan couple, and when setting out for the 'final sprint' today, our heavy ruck-sacks stay behind with them. There are only a few items we need to take with us: some drinks, rain clothes and a package of biscuits, just in case… Our host lends us his old, torn day pack for the remaining kilometres.

This last day, which I have been looking forward to for such a long time, starts with me plodding along the dirt road that leads to the Nara La (4,560 metres) the last pass before the border. The map shows two, but one of them is nothing more than a big, flat spot on the way down to Hilsa.

It comes as a relief when we, finally, get to a small trail offering the chance to get away from the road. This saves us some time, though the extreme steepness of the path slows us down considerably. We cut the road a couple of times and two hours later arrive at the pass, which provides an unobscured view of the vastness of the Tibetan high plateau. The infinite emptiness evokes a strong desire to walk on forever. At least, I would like to continue to Mount Kailash; the sacred mountain. Of course, due to Chinese laws and regulations, it is impossible for individuals to obtain a permit for a trip like this and, all of a sudden, sadness grips me and I plunge into a dark inner void.

I am still in a depressed mood when entering the village of Hilsa one hour later. What about the joy I had expected to fill my heart? What about feeling proud? I have successfully completed a four-month

Our hosts

journey across Nepal. There should be feelings and emotions. Nothing, nothing at all. All I experience deep inside me are growing desolation, forsakenness and despair. I stroll through the settlement in search of something that may evoke joy or happiness, but I find the atmosphere just as gloomy and depressing. I walk over to the bridge and attach the khata, which I got from Pimba's mother many weeks ago, to the bridge railing and sit down on a stone wall. From here, I watch Chinese trucks arriving at an iron gate on the other side of the river and the endless stream of porters carrying the goods from the gate over the bridge to Nepal. It is mainly cheap alcohol.

Successfully, I fake a bright smile when Temba takes a couple of pictures of me, but I do not feel like laughing; not at all. It feels as if the meaning of my existence has evaporated. I suppose that in one way or other this is correct because for the last eight months all my thoughts and dreams circled around this journey. Now, this very moment, it is over and I fall into a deep, black abyss. What can I fill my dreams and thoughts with in the future? Right now, I have no idea.

We have lunch in a restaurant that is part of the local police station and the border control. In the early afternoon, our driver arrives, and we return to Yari.

Three days later we are back in Simikot, and the following morning we sit on a plane to Kathmandu.

Postscript

In the introduction I wrote that the Great Himalaya Trail was to be my 'last great adventure in life'. But plans are there to be changed and this is no exception…

Since I began my walk, many things in my life have changed. I no longer work – or wish to work – in a 'normal', full time job. Instead, I have begun to work on long-held ambitions and projects: re-discovering my love of photography, writing this book and developing a related stage-lecture – and promoting Autism through these channels. I see now that there are only self-constructed limitations in life and have resolved to try to do the things I really want to do. As I write this, I am about to set off on another trip – another 'little madness' – my 'Seven Summits of the Alps' tour. Again looking to raise awareness of Autism, I will visit the highest summit in each of the seven countries of the Alps, starting in Slovenia and travelling from mountain to mountain by bike.

Who knows what will come after that…

For information about lectures and new adventures, please visit www.gerdapauler.info

ACKNOWLEDGEMENTS

A special thanks goes to Sir Chris Bonington for taking on the role as the patron of my Charity Walk Project and for writing the foreword.

Many things could have gone wrong – totally wrong. The reason why this did not happen is the sole merit of my Nepalese friends and helpers:

Temba Bhoti, my guide, who did an extraordinary job, and whose open hearted mind helped me to cope with the hardships of a four-month trip.

Lakpa Bhoti, my cook, who looked after me for many weeks. He prevented me from losing too much weight and looking like a bag of bones.

Pimba Bhoti, Sonam Bhoti, Jomma Bhoti and Mingma Bhoti, who carried our equipment across the country, stepped in as cooks when necessary, served tea and coffee and who radiated so much joy. Their happiness was infectious and helpful to me.

Kinsang, Mingma Bhoti, who joined me for a short period of time but even so they contributed to the success of my trip.

Thanks to my sponsors who not only contributed gear for me but also for my Nepalese group so my travel budget did not get out of control:

Atello, Beal, Bergans of Norway, Bridgedale, Buff, Kaahtola, La Sportiva, Petzl.

Nilam, my agent from Asian Heritage Treks and Expedition, who fixed visas and permits

Ulrich Schroeder and Robin Boustead for allowing me to use their maps and pictures free of charge for this book and my lectures.

Uttam Phuyal at Kathmandu Guesthouse, who took on the role of looking after the donations and for always having a free room for me at the hotel.

David, Mountain People, who was there to sort out many things and who contributed a lot to the success of the charity project. He made me understand and appreciate honest friendship.

FURTHER READING

Guidebooks

Nepal Trekking and the Great Himalaya Trail, Robin Boustead.

Trekking in the Nepal Himalaya, Stan Armington. Eighth edition of the most complete guide to trekking routes in Nepal.

Lonely Planet Nepal – 5th edition, Tony Wheeler and Hugh Finlay. A complete guidebook to Nepal.

Trekking in Nepal: A Traveler's Guide, Stephen Bezruchka. Has cross-cultural suggestions and detailed information about how to organise a backpacking or lodge trek. There are many route descriptions that include walking times.

The Trekking Peaks of Nepal, Bill O'Conner.

Mount Everest National Park: Sagarmatha Mother of the Universe, Margaret Jefferies.

Guide to Royal Chitwan National Park, Margaret Jefferies and Hemanta Mishra.

Trekking in the Everest Region, Jamie McGuinness.

Trekking in the Annapurna Region, Bryn Thomas.

Mountains, Mountaineering & Fiction

Into Thin Air: A Personal Account of the Mount Everest Disaster, Jon Krakauer.

Travelers Tales – Nepal Stories and anecdotes by those who have lived in and visited Nepal, including Jimmy Carter, Jeff Greenwald and Broughton Coburn.

The Seven Mountain Travel Books by H W Tilman which contains *Nepal Himalaya*. A delightful book filled with Tilman's dry wit which describes the first treks in Nepal in 1949 and 1950.

Escape from Kathmandu, Kim Stanley Robinson. An off-the-wall romp around Nepal. There's an effort to free a captured yeti, an illegal ascent of Everest with a reincarnate lama and an encounter with Jimmy Carter.

The Ascent of Rum Doodle, W E Bowman. The classic spoof of mountaineering books. It's a good diversion after reading a few expedition accounts that take themselves too seriously.

Video Night in Kathmandu: And Other Reports from the Not-So-Far East, Pico Iyer.

Seven Years in Tibet, Heinrich Harrer.

Stones of Silence: Journeys in the Himalaya, George B. Schaller.

East of Lo Monthang: In the Land of Mustang, Peter Matthiessen and Thomas Laird.

The Snow Leopard, Peter Matthiessen.

Karnali Under Stress Livelihood Strategies and Seasonal Rhythms in a Changing Nepal Himalaya, Barry C. Bishop.

Trans-Himalayan Traders: Economy, Society, and Culture in Northwest Nepal, James F. Fisher.

Nepali Aama: Life Lessons of a Himalayan Woman, Broughton Coburn. An account of a US Peace Corps volunteer's encounters with an elderly Gurung woman. Coburn later took her on a trip to the US; this trip is described in a sequel, *Aama in America: A Pilgrimage of the Heart.*

Sherpas: Reflections on Change in Himalayan Nepal, James F. Fisher.

Storms of Silence, Joe Simpson.

Tiger for Breakfast: The Story of Boris of Kathmandu, Michel Peissel.

Penguins on Everest, David Durkan.